I0793681

IMAGES
*of America*

# SHERIDAN AND GRANT COUNTY

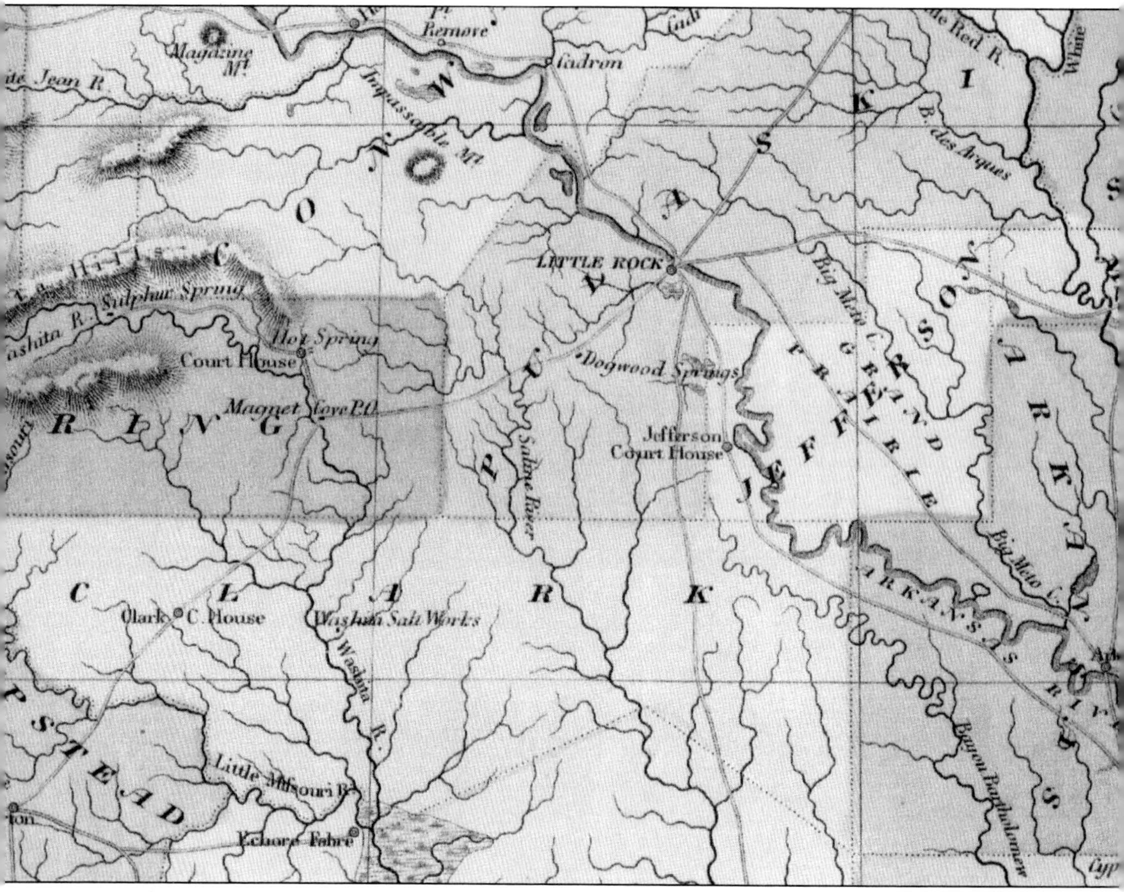


Southward, a stagecoach road joining Little Rock with Louisiana, appears on an Arkansas map of 1835, when Pulaski and Clark Counties governed what is now Grant County. (Courtesy of Special Collections, University of Arkansas Libraries, Fayetteville.)

**On the Cover:** A century ago, mule power and hand tools assisted bridge construction in southern Grant County. Seated from left to right are Miller, Scales, crew boss Stubbs, and Mack Ashcraft; Dan Atchley stands in the log in the background under Berry Clark and Perry (holding ax); Van Stubbs (hand on keg), Judd Gwin (arms crossed), and Lawrence Hamilton (hands folded) stand in front of, from left to right, Gabe Wilson, Willie Moore, Newt Street, Dink Wilson, two unidentified, and Cliff Moore. (Courtesy of the author.)

IMAGES
*of America*

# SHERIDAN AND GRANT COUNTY

Roy L. Wilson

ARCADIA
PUBLISHING

ISBN 978-0-7385-9439-2

Published by Arcadia Publishing
Charleston, South Carolina

Printed in the United States of America

Library of Congress Control Number: 2012943293

For all general information, please contact Arcadia Publishing:
Telephone 843-853-2070
Fax 843-853-0044
E-mail sales@arcadiapublishing.com
For customer service and orders:
Toll-Free 1-888-313-2665

Visit us on the Internet at www.arcadiapublishing.com

*This book is dedicated to my parents, Boe and Connie Wilson, born in Grant County and members of the Greatest Generation, who lived the transition from horse-and-wagon days to modern Arkansas.*

# Contents

# ACKNOWLEDGMENTS

Education, when embraced, has the power to enrich lives. This book salutes the centennial of the 1913 Sheridan school building and the role it has played in fostering progress, fulfilling the admirable goals of education. Thank you to the Sheridan School District for the four decades of support of my teaching career. To my former students: your encouragement and compliments over the years sustained me and guided me toward a deeper understanding of education and especially of the history of Sheridan and Grant County.

I am especially grateful to Lindsey Allen Stanton of the Grant County Museum for her kind assistance and encouragement; D.J. Wallace, museum director, for supporting this effort; Hazel Allen Wilson for her guidance and wisdom; and Grant County and Circuit Clerk Carol Ewing for her confidence in my undertaking this project. The scrapbook created by Retha Jones Johnson, longtime educator and native of Sheridan, documented her high school years from 1927 to 1931 and captured the culture of the time. Thanks to her grandson Doug Lloyd for permitting access to her collection.

Thanks to all who helped me, especially Benny Barnes, Dr. Curtis and Lois Lea Clark, William Hobart Byrd Jr., Jim and Betsy Taylor, Joan Crowson Burton, Shirley Coppock, Jim and Marlo B. Krueger of the Saline County History and Heritage Society, Dr. Bill J. Gurley, Arthur Maune, Rick Meadows of the Civil War Roundtable of Arkansas, Leon and Elizabeth Ann Lancaster, Patty Vance Hays, Lessie Wallace, Bob Bratton, Peggy Reynolds, Virgie DuVall, Linda Atkins, Alicia Clement, Grant County Library, Jimmie Koon, Charles Boyd's *At Liberty on Bear Creek*, Mike Howard and Bill Prior of the Arkansas Geological Survey, the Army National Guard Museum of Arkansas, the Old State House Museum, Arkansas History Commission, Butler Center for Arkansas Studies, and Special Collections, University of Arkansas Libraries, Fayetteville. Thanks also to Simone Monet-Williams and Mike Litchfield of Arcadia Publishing and the large number of Grant County relatives and acquaintances over the years who willingly shared their stories.

# Introduction

Gravels and soil of Sheridan and Grant County date from the Pleistocene epoch. Significant flooding and large fluctuations in climate occurred during ice ages from 2.6 million to 12,000 years ago. Gravels from high terraces along the Saline River and the county's northeast washed in from the Ouachita Mountains during the Quaternary period. Silica from flood events petrified buried trees. Red clay, lignite deposits, and gravel fossils date from the Eocene period of 45 million years ago. Some gravel shattered from much older rock formations. Chemicals in water naturally cemented rocks and gravel into conglomerate. The ancient Sparta aquifer, 600 feet underground, supplies Sheridan's water.

Native Americans traveled from the Arkadelphia area toward Pine Bluff as late as the early 1800s, passing through one of their large settlements at a Saline River crossing named Jenkins' Ferry in the 1840s. New settlers chose land near the Saline; the earliest known, William Lockhart, claimed land in 1815 four miles southwest of Benton at a river crossing. By 1819, ten families arrived. Saline County, named after an early saltworks, contained most of Grant County from 1835 to 1869, with Pulaski, Clark, Arkansas, Hot Spring, and Jefferson Counties previously sharing jurisdiction.

A depressed economy in 1837 hurt the Hagler and Bear Creek communities of Bibb County, Alabama. Elder Joab Pratt, moderator of their Tuscaloosa Baptist Association in 1839, led a wagon train in late 1841 with the families of George and Azariah Cobb, A.J. Mayfield, Nathan Pumphrey and his wife, Louisa, her father, Berryman McDaniel, the Brazeal family, and others "to a place in old Saline County, Arkansas, where they founded Philadelphia Baptist Church and a community called Prattsville," according to *At Liberty on Bear Creek*. To the west was Pratt's spring, the county's largest known spring flow.

Abundant wildlife and stands of virgin oak supplied these hardy families. The prosperous 1850s saw the arrival of two of this area's wealthiest families, Alexander and Jane White Pinchback from Rock Hill, South Carolina, and Dr. Richard Clinton and Susan Davis Rhodes of Clinton, North Carolina. Within a decade, tragedies would befall both families that money could not prevent.

To ease wagon travel to visit courthouses, legislation during Reconstruction established a new county and county seat named for Union generals Grant and Sheridan. Blacksmith shops, two general stores, a church, and a post office were soon situated along impassable dirt roads. Survival skills were shared across generations. Electricity arrived in rural Grant County 50 years after installation in the White House.

In 1903, Grant County Bank was the first bank established in Sheridan. The State Bank of Leola closed during the 1930s. The early 20th century saw natural gas replacing wood stoves in town. Farm income depended on cotton prices. In 1909, the third courthouse was constructed of brick by commissioners J.W. Donham, J.B. Shaw, and W.J. Nixon. Monk and Ritchie were the contractors. In 1913, they also built the county's first brick schoolhouse in Sheridan. A century later, the Sheridan School District is the county's largest employer and continues the tradition of young men and women seeing their families to a better future through education.

World War I took several local men on their greatest adventure. The Grant County War Memorial honors 10 of these men who were casualties of that war. Several county residents died during the 1918 flu epidemic. The Missionary Baptist College was constructed in 1921. A few homes received electricity in 1922 from Sheridan Light and Power Company generators. In 1929, a water and sewer system was established. When the J.L. Williams and Sons sawmill arrived, incomes improved. Almost every employed adult, if not farming, either cut and hauled timber or toiled in county sawmills. The Great Depression offered the same life as before—hard work and little money. At least 46 county men would die in World War II, two in Korea, and seven in Vietnam. Families prospered by operating service stations and retail stores. In 1947, new concrete bridges spanned swollen creeks. Trains moved Sheridan passengers until the 1950s and lumber until the 1980s. In the 1950s and 1960s, companies Arkla Meter Repair, Rockwell International, and West Bend arrived. Sheridan enjoyed several modern church buildings, a new Masonic lodge, paved streets, a tennis court, a swimming pool, and a library, the latter made possible by a donation from Dr. and Mrs. O.R. Kelly.

Some residents recall an era when youngsters made toy cars with jar lid wheels, competed for marble championships, had to work 40-acre fields, and picked cotton in the fall. Adults gathered hickory nuts and pecans for winter treats, carried things in tow sacks and lard buckets, and collected hay for mattresses to support feather beds. They attended whole hog pit barbecues, went to summer revivals with baptizing in ponds or the Saline River, built their own houses from trees sawn at a nearby mill, and saw older people using turkey tail fans to stir up a breeze. Blue tick and black and tan dogs roamed yards, lucky to get an old biscuit thrown out the back door. Cats were tolerated only because they caught mice. Families ate turnip greens, pinto beans, and corn bread as staples, churned butter, cooked possums out of necessity, pinned hunting licenses to hats, and received driver's licenses, simply handed out by officials, for driving on roads so dusty they discolored nearby trees.

Ladies crafted homemade clothes and quilts from flour and fertilizer sacks. Paying poll taxes was a requirement to vote. Men on road district crews cleared rights of way in lieu of taxes. There were no school cafeterias, basketball courts were dirt, community sports teams welcomed anyone who wanted to play, and nighttime football games were lit by car lights facing the field. Fistfights between brawny farm boys occurred at the slightest provocation, and 17-year-olds drove school buses. Battery-powered radios brought sounds from faraway places. Medicine was scarce, hookworm infestations were common, kitchen sinks drained into yards, and outhouses were accompanied with wasp nest worries.

Providing food trumped everything else. Community adults were expected to publicly correct misbehaving children. When television arrived, 20 people might gather to catch a glimpse. Hot summer nights found sleep impossible, even with electric fans. Chaperones were in vogue at all social events. Entire families gathered on porches for hours, with guests spending the night on quilts spread on the floor. Water was drawn from the well, washtubs were bathtubs, windows had no screens, and chills occurred from malaria-infested mosquitoes. People would share their garden's bounty, reputations were guarded as if gold, and people would dress up to go to town. Stores closed on weekday afternoons, and blue laws prevented selling anything on Sundays. Preachers and fried chicken were obvious and welcome at Sunday dinner, coal oil lamps expelled the blackest dark, bedbugs endured, and crackers and flour were sold in barrels.

Bare dirt lawns tracked snakes entering yards, corn-shuck mops and sand cleaned floors, dogwood branches became yard brooms, and twigs became toothbrushes for baking soda and salt toothpaste. Lime pickles were canned, blackberries picked, and the Fourth of July was celebrated with homemade ice cream. For 25¢, people bought a Coke, a candy bar, and a movie ticket. Business was conducted by handshake, because a man's word was his bond. Country chewing gum was sap oozing from a sweet gum tree. Stores mostly sold bolts of cloth ready for mothers to sew, and every father knew wooden homemade boats were made for paddling on lazy summer days.

When all of these things are understood, people will know how Sheridan and Grant County came to be.

# One

# Ancestral Beginnings

State land records witness the arrival of a new era as the final Quapaws, Cherokees, and Choctaws signed over land, despite their belief of "shared birthright," and acquiesced to the settlers' idea of deeded ownership. Among names on local deeds of the 1850s were: Abbe Ko Che Ak, Hah O Le Tim Ah, Te Ho Yo, Fish O Ham Be, E En Lah Tubbee, E Mi Ukla, and Hik E Yah. An 1823 Arkansas map labeled a section "Quapaws," along 34 degrees latitude from the Mississippi River across the Saline to the Ouachitas. The first territorial map of Arkansas lists "Indian Territory" from west of Arkansas Post to the Saline River and north to the Arkansas River. Clark County, 10 years old at the time of the map, began on the west bank of the Saline. Arkansas County extended well into present-day southeastern Grant County. In 1835, Pulaski County was the northern half of Grant, and Clark County the southern.

Early purchasers of land in February 1834 along the Saline River in township 2 south, range 15 west, north of present-day Tull, were Arden and Samuel Pelton. On April 15, 1837, in township 4 south, Richard Cocke bought 3,200 acres; John C. Cooper purchased 1,500 acres in township 5 south, west of Prattsville. On the same day, Jonathan P. Sneed purchased 640 acres in section 36, township 6 south, range 15 west, southeast of present-day Leola. On July 28, 1838, a wealthy speculator and lawyer from Vicksburg, Mississippi, William S. Bodley, purchased 2,943 acres in township 6 south, range 15 west, located a few miles northwest of present-day Leola.

On November 11, 1848, Abraham Allee purchased 80 acres in section 34, township 6 south, range 13 west, southwest of the Shiloh community, and Joseph Berry bought 160 acres in section 19, west of Shiloh. Thomas H. Bradshaw purchased land from Edward Calvert as hundreds more families arrived in the 1850s and 1860s in what was then Saline County.

These ancestors were in touch with nature's time, knowing when to plant and lay by crops; the time to harvest, preserve food for the winter, save seeds for next spring's planting; the time to hunt, fish, gather, and slaughter to fill a smokehouse; the time to get right with the Almighty; and the time to say goodbye, since death could arrive most any day. Land was the dream that provided shelter, food, and a livelihood. Without hard labor and discipline, our ancestors' survival was not assured. They gave future descendants who cared the opportunity to reach toward unimagined dreams.

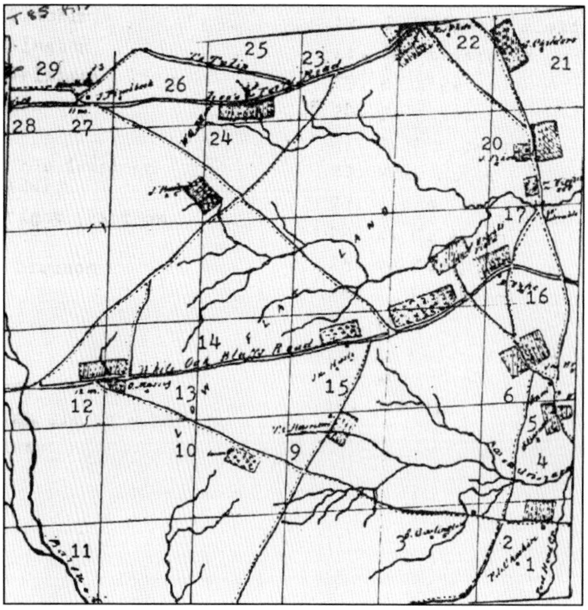


In the 1840s and 1850s, a prosperous community of progressive families existed south of the Saline River along the Lee's Ferry and White Oak Bluffs Roads. This captured Confederate map detailed, by number, farmsteads in township 8 south, range 13 west: DeKalb Wylie (16), William G. Allen (24), Jane Pinchback (26), sawmill (29), and Spirons Rogers (28). This section transferred from Dallas to Cleveland County in 1873. (Courtesy of Arkansas History Commission.)

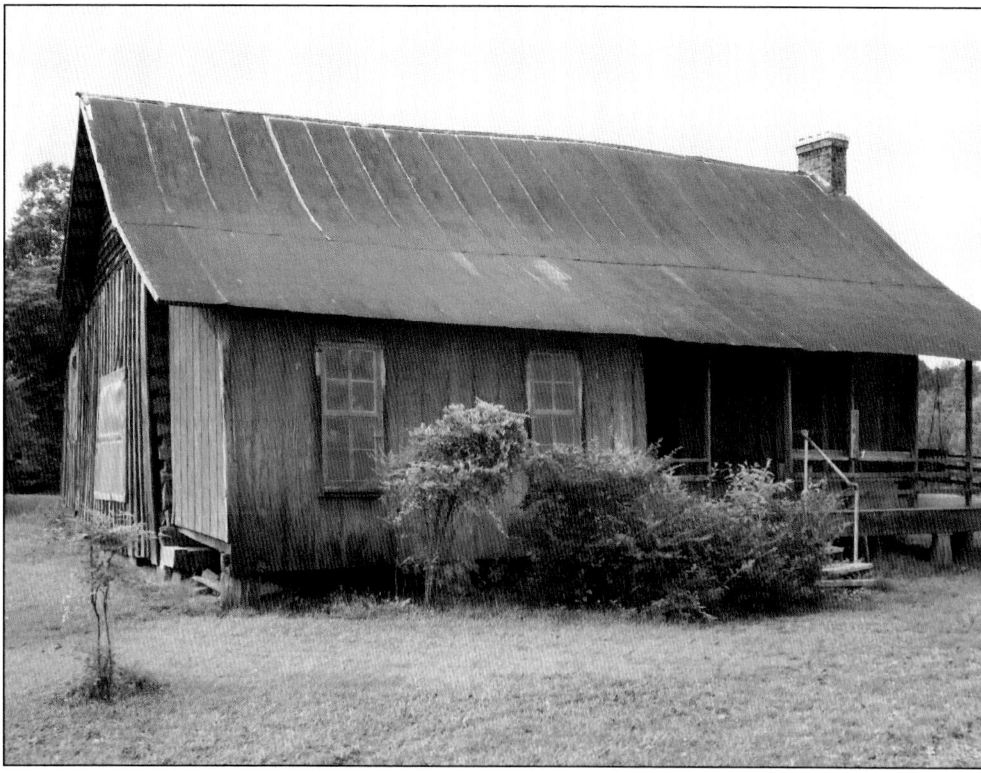

In 1848, Samuel Daniel Byrd of Tennessee settled on the Pine Bluff to Rockport Road and constructed a cabin (logs visible at left), followed by another room to the east in 1850 separated by an eight-foot breezeway. Additional rooms were added in 1896. Originally in Fenter Township, Hot Spring County, it is one of the oldest structures remaining in Grant County. (Courtesy of W.H. Byrd Jr. family.)

Tabitha Jane Reed of Tennessee married Samuel Daniel Byrd in 1850. Their children were Lovey A., William S., Mary M., Susan C., Matilda Roxie, James C., and Samuel Daniel Jr. During the Civil War, a Union soldier approached their house, whistled to get attention, then asked for some food. (Courtesy of W.H. Byrd Jr. family.)

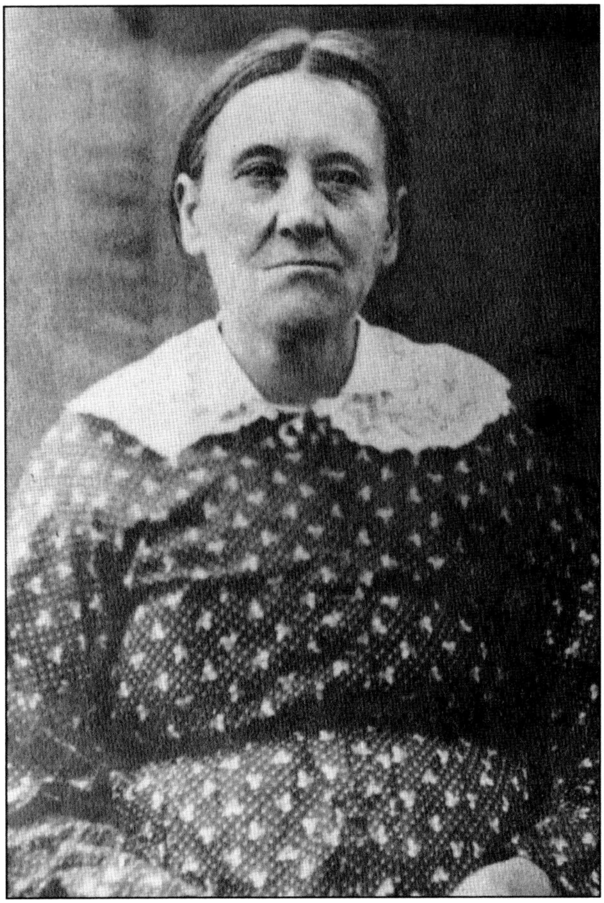

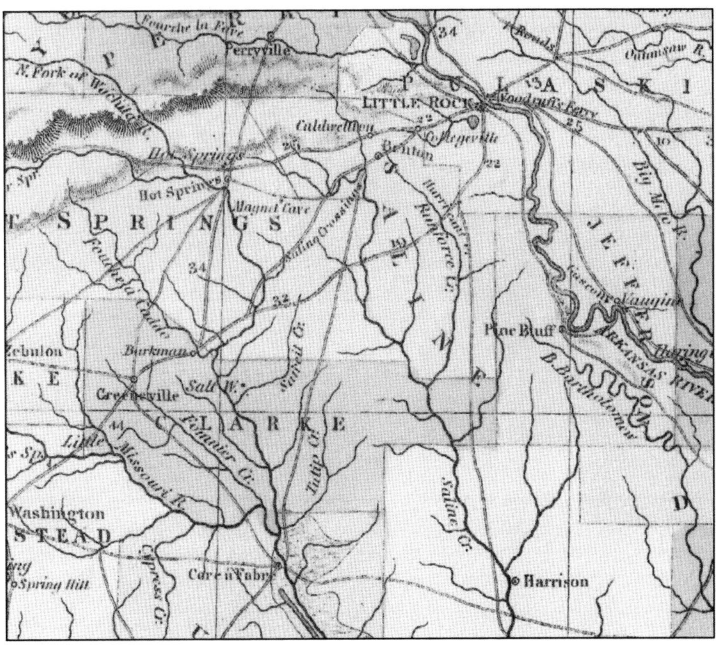


This map, published in 1846 as "A New Map of Arkansas" by S. Augustus Mitchell in Philadelphia, shows additional stagecoach roads joining Core a Fabre (Camden) with Little Rock through Saline, Hot Spring, and Clark Counties. The name "Runforce Creek" (Hurricane Creek) disappeared. (Courtesy of Arkansas History Commission.)

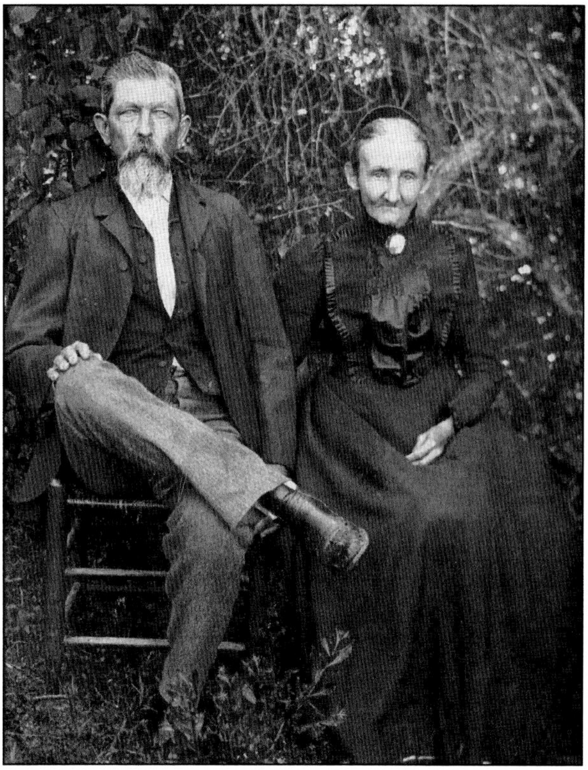

John Rufus Taylor, born in 1839 in North Carolina, moved at age 15 with the Riggan family to settle 160 acres southwest of the Saline River in what was then Hot Spring County, north of present-day Leola. He married Sarah Riggan, who inherited another 160 acres. On August 1, 1861, Rufus joined Company I, 12th Arkansas Infantry, then Company F, 3rd Trans-Mississippi, Morgan's Regiment, and the 26th Regiment Arkansas Infantry, CSA. He attended the Civil War reunion in Little Rock in 1911, donated land for the Taylor Cemetery, and died in 1927. (Courtesy of Jim and Betsy Taylor.)

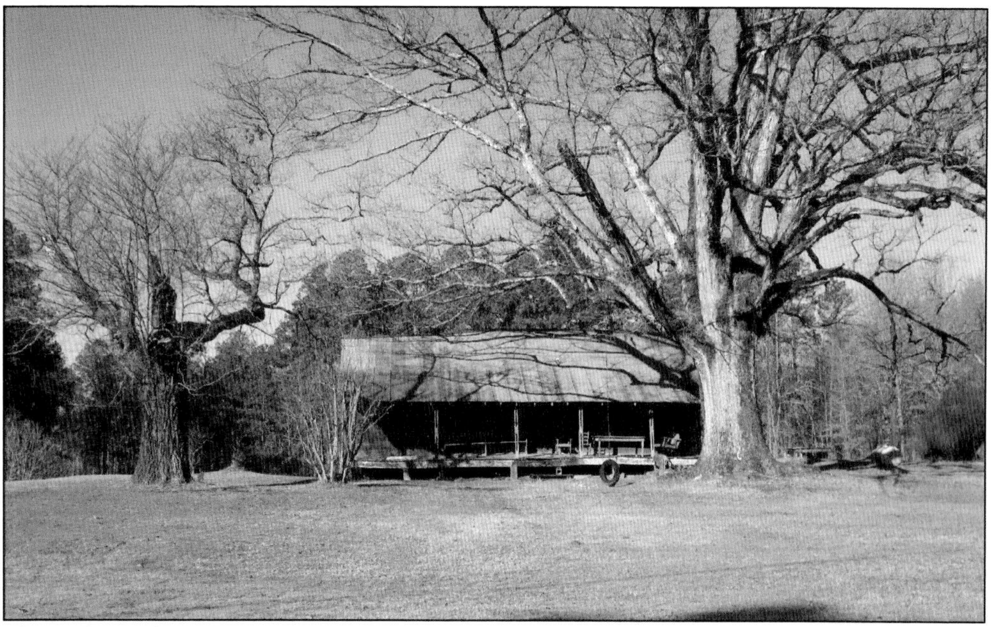

The Rufus Taylor House is one of the oldest structures remaining in Grant County. Originally a one-room log home built in the 1850s on the same property, it was moved to its present location and expanded into a dogtrot. The large white oak tree at right was already providing shade during the Civil War. The tree to the left is one of the largest sassafras trees in Arkansas. (Courtesy of the author.)

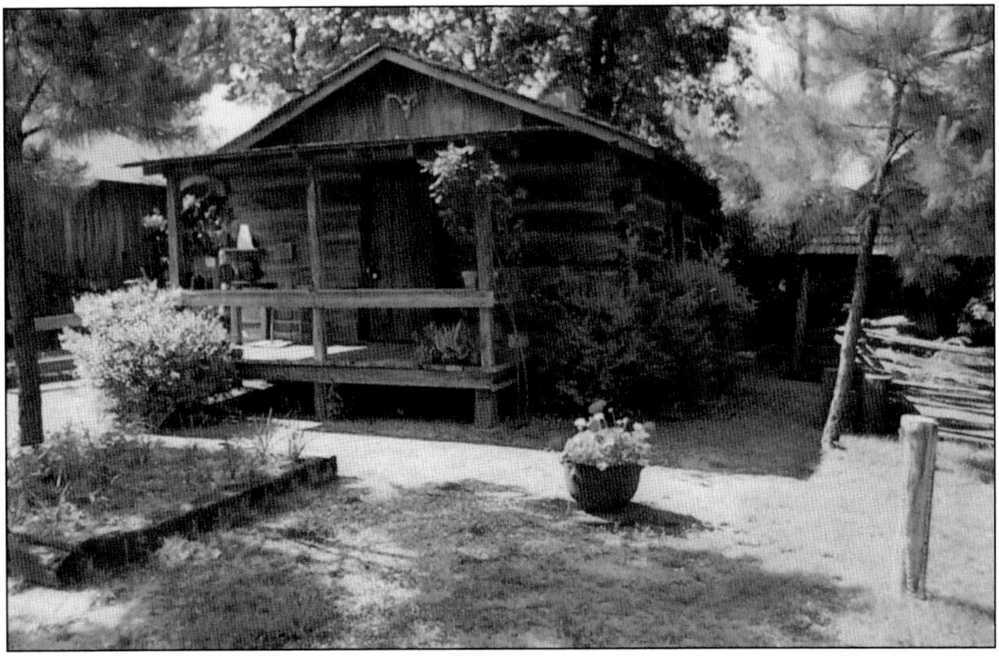

James Bratton, of Brattonsville, South Carolina, arrived in the late 1840s to build this cabin northeast of the present-day Crossroads community, then in Saline County. The structure was about 25 feet square. Bratton's initials were carved into an outside log. The building was donated and once exhibited at the Highway 270 location of the Grant County Museum. (Courtesy of Grant County Museum.)

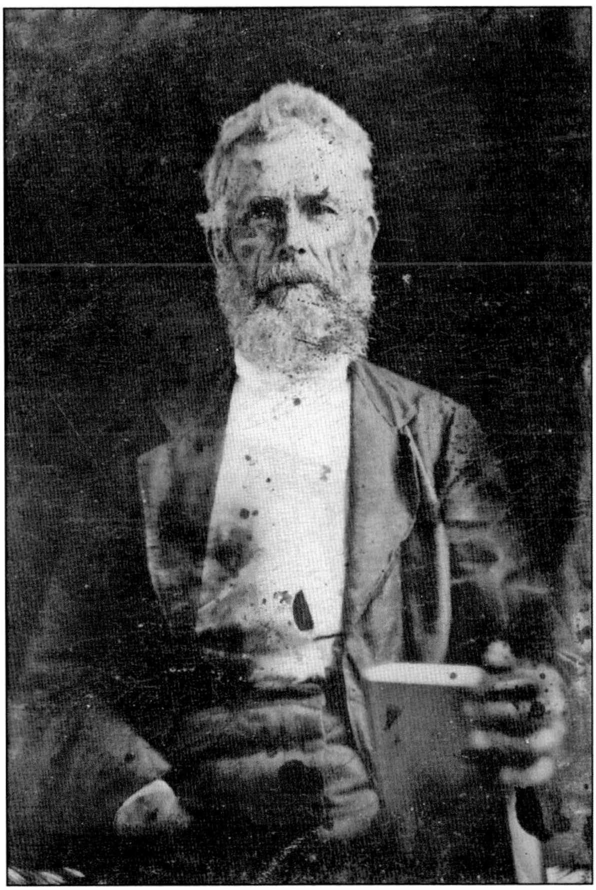

Joel Rushing was born in Bedford County, Tennessee, in 1823, the son of John and "Millie" Amelia Deason Rushing. He married Nettie Caroline Trawick (born 1821 in Alabama) in 1846 in Tuscaloosa County, Alabama, before settling in what was then Saline County. Rushing was one of the founders of Lost Creek Baptist Church in Sheridan. (Courtesy of Lois Lea Clark.)

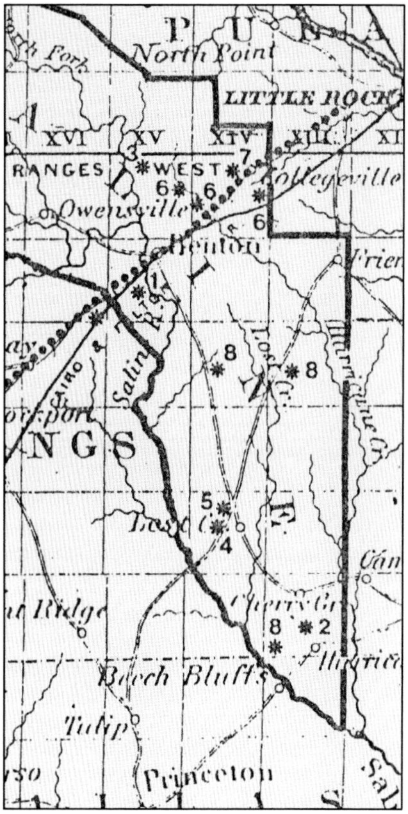


An 1850 map entitled "Location of Lands of Top Cotton Producers and Slave Owners in Saline County, Arkansas" lists cotton bales produced in what later became Grant County: Edward Calvert (farm 2), south of Cherry Grove, 38 bales/17 slaves; Berryman McDaniel (4), 30 bales/17 slaves; George Cobb (5), 20 bales/15 slaves; and John S. Gee (8), 50 bales/11 slaves. Others listed were Abram Allee, Seth Atchley, William Crowson, James Gamble Sr., R.B. Gamble, J. Rodgers, and Jesse and Elizabeth Reeves. (Courtesy of Arkansas History Commission.)

Alley's Tannery and Peacock's Grist Mill were in business in the 1850s beside an early road in section 32, township 5 south, range 13 west, just across the Saline River from Big Island. Alley's reported harassment from Union patrols during the Civil War. Cypress-shingled buildings lasted until the 1920s. A hill 190 feet above sea level is the site of Alley Graveyard, one of the oldest cemeteries in the county, which served families with surnames including Alley, Gamble, Pelton, and Allen. During the late 20th century, the site was used as a deer camp. (Courtesy of *The Official Military Atlas of the Civil War*, Barnes & Noble Publishing.)

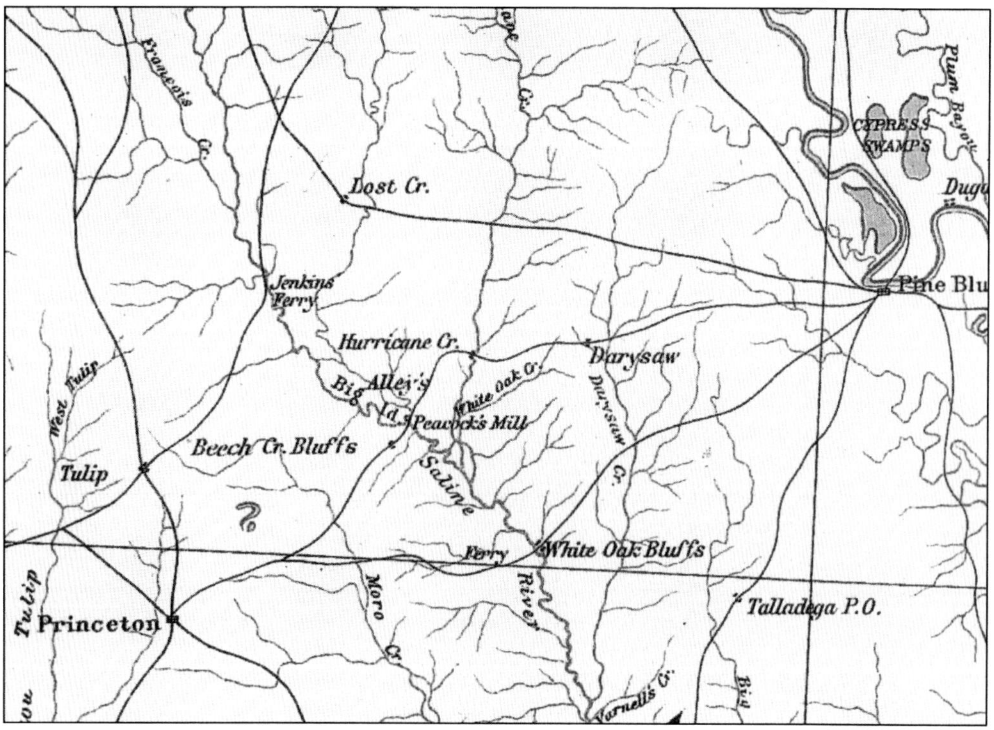

In 1858, Isaac Brumbelow settled on 240 acres west of the community of Crossroads, then located in Hot Spring County at the junction of several early roads. In the early 1900s, the town was renamed Poyen. Shown here is a tintype of family member John Brumbelow. (Courtesy of W.H. Byrd Jr. family.)

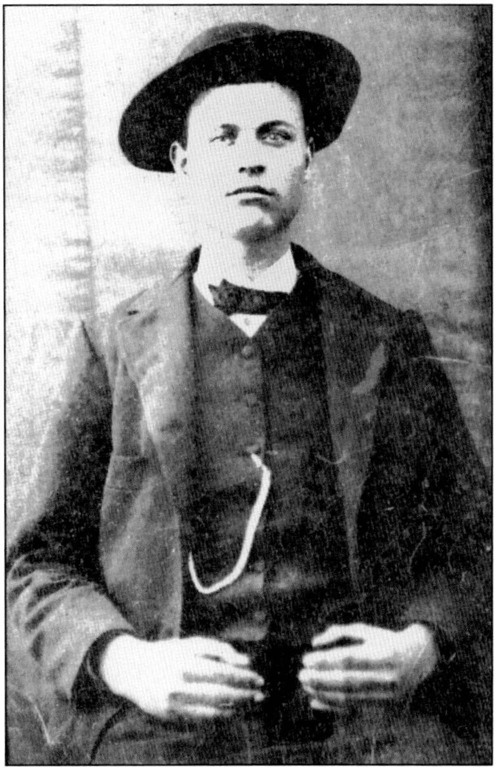

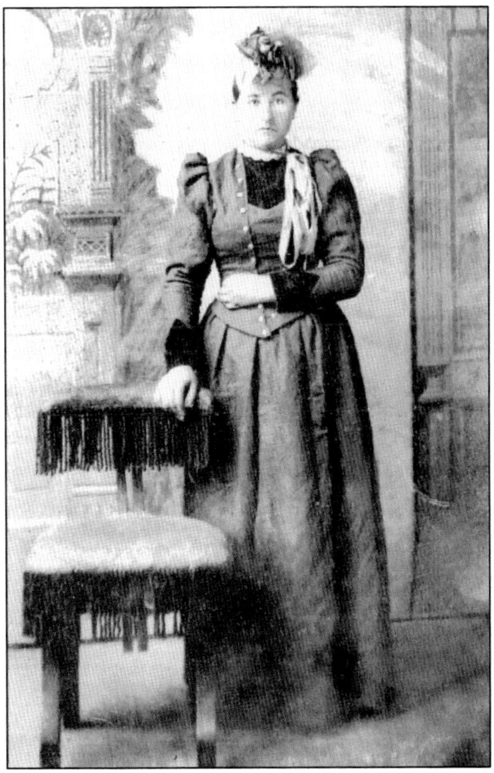

Mary Ann Rowland was a member of a family who were early settlers in the Poyen area. Jeremiah Rowland homesteaded 240 acres in 1860. Earlier settler Calvin Allred purchased land near Poyen on August 2, 1838, the same day Henry Bennett did so, in the vicinity of Tull. (Courtesy of W.H. Byrd Jr. family.)

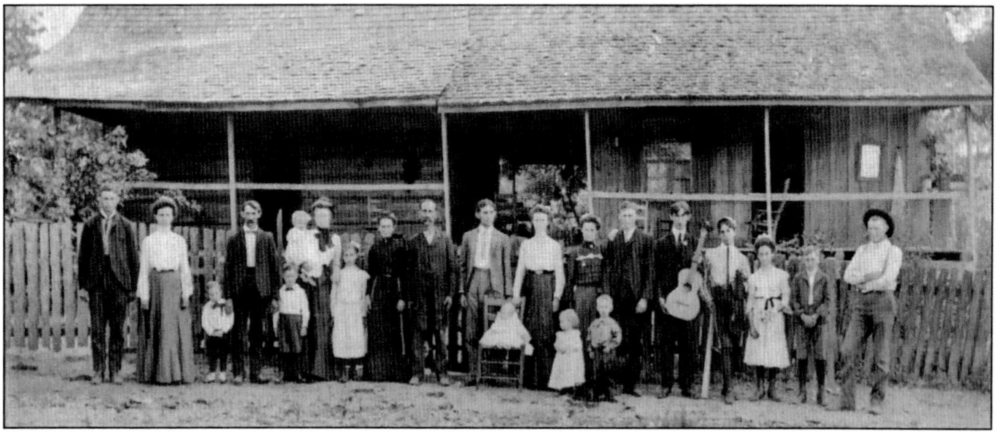

The Tull community's progressive DuVall family poses for a photograph. The family members are, from left to right, Thomas Brumbelow, Nora DuVall Brumbelow, Alvin DuVall, Will DuVall, Ira DuVall, Lula Jacobs DuVall (holding Essie DuVall), Mae DuVall, Elizabeth Grimmet DuVall, George A. DuVall, Elisha DuVall, Wilbur DuVall (on chair), Ida Jacobs DuVall, Rose DuVall Jacobs, Frank Jacobs (with children Ida Belle Jacobs and Carthel Jacobs), George DuVall (with guitar), Vernon DuVall (with violin), Mattie DuVall, Alfred DuVall, and Lee Jacobs (brother of Frank Jacobs and Lula and Ida Jacobs DuVall). The house was built before the Civil War by Samson Grimmet, who was killed at the Battle of Chickamauga. (Courtesy of Virgie DuVall and Joan Crowson Burton.)

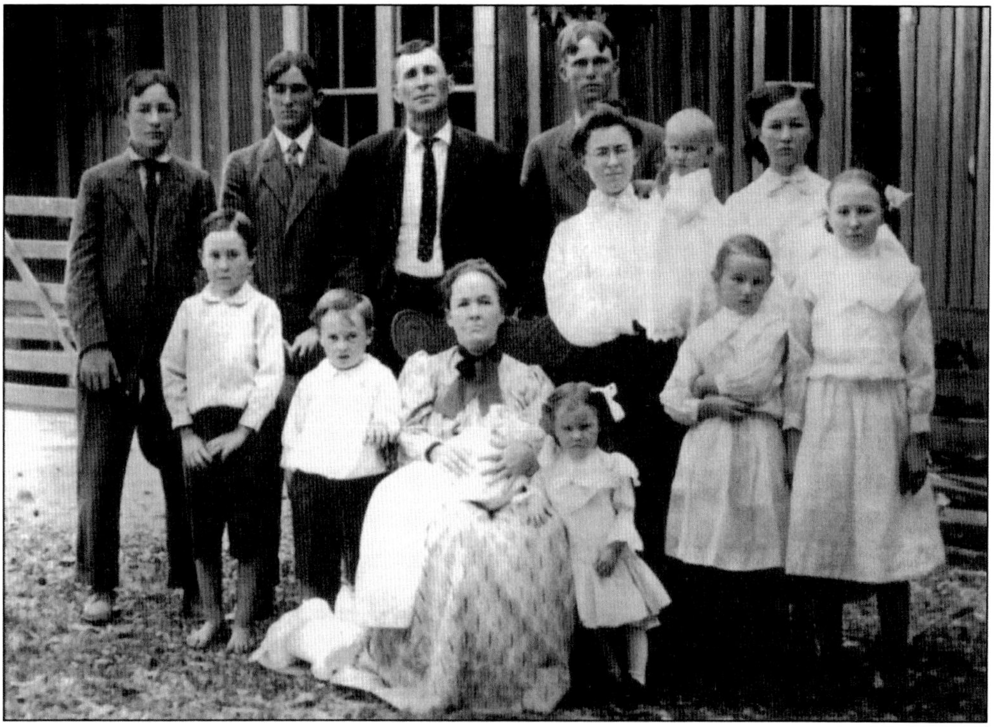

Residents of the small town of Tull shown here are, from left to right, as follows: (first row) Gus Tull, Herbert Tull, Susie Scarbrough Tull (holding Gertrude Tull), Alma Smith, Mittie Smith, and Angie Tull; (second row) Luther Tull, Fred Tull, Dow Tull, Henry Hambrick, Alpha Tull Smith (holding Richard Smith), and Annie Tull. (Courtesy of Virgie DuVall and Joan Crowson Burton.)

Emma Sweeney, daughter of Samuel D. Byrd Sr., married Tom Sweeney but unfortunately died of complications during childbirth. (Courtesy of W.H. Byrd Jr. family.)

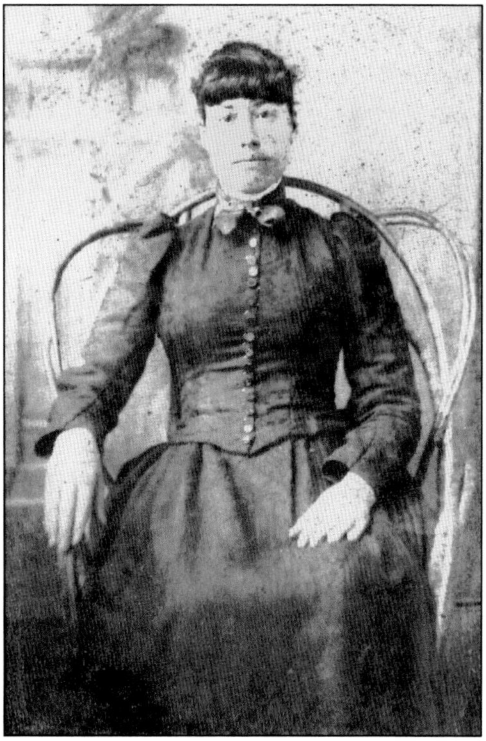

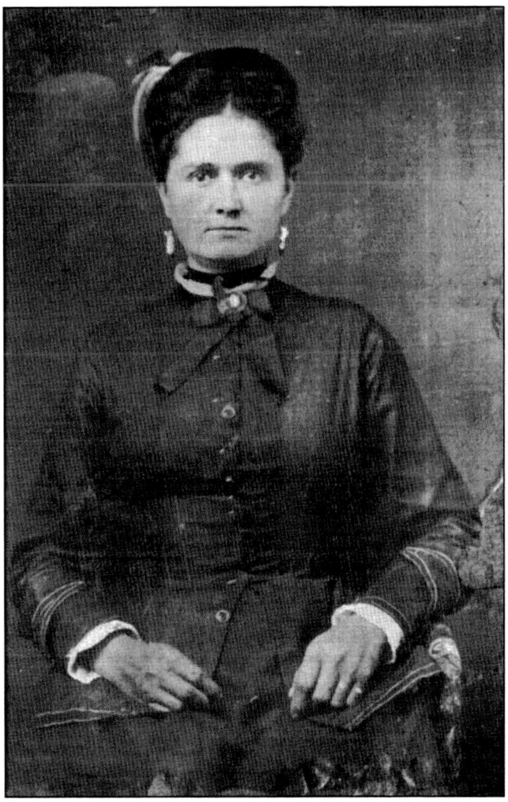

Emily Brumbelow, sister of John Brumbelow, married Jim Byrd, son of Samuel Daniel Byrd Sr. of Poyen. Earlier in the 1830s along the Saline River, settlers included Eli Lockhart, George Keesee, Joseph Lawrence, John Cooper, George W. Murphy, Andrew R. Govan, Benjamin Shaw, John S. Skinner, and relatives James, John and Theophilus Ratcliff. (Courtesy of W.H. Byrd Jr. family.)

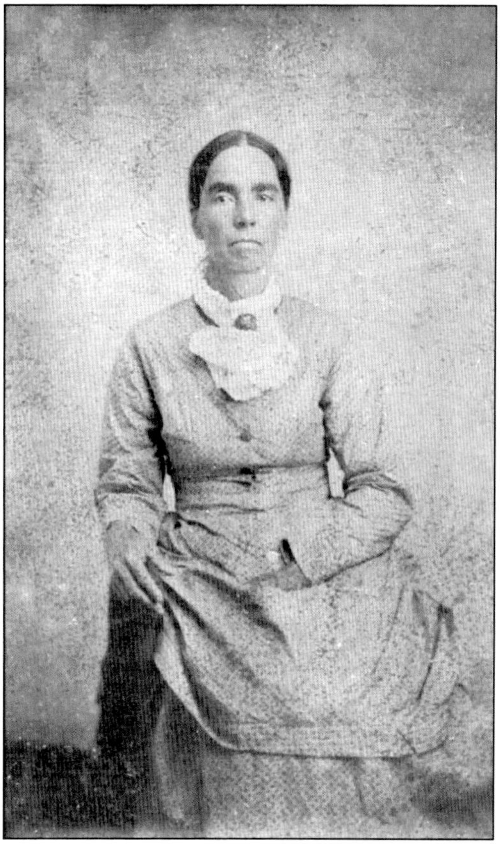

For local people arriving at adulthood before 1860, troubles beyond description awaited to disrupt their lives, families, and fortunes. The unidentified woman in this tintype assumes a determined look, hoping for the best. (Courtesy of Grant County Museum.)

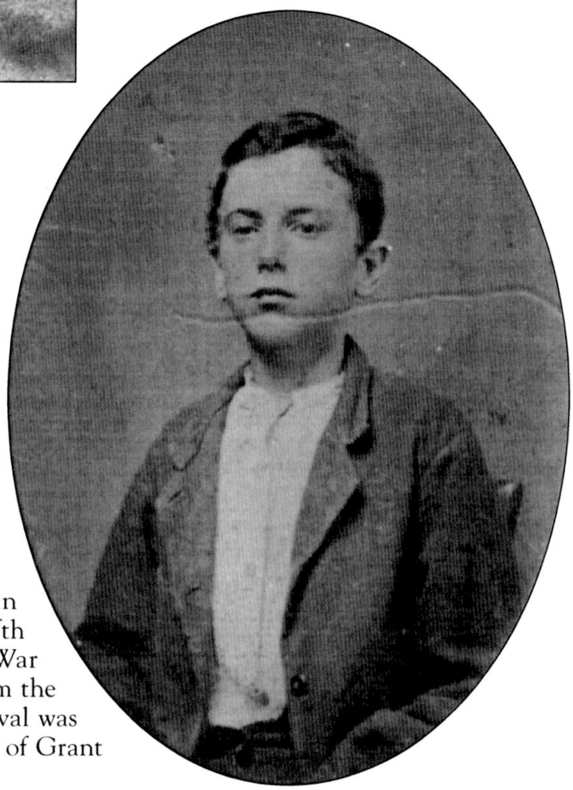

Until the 1920s, half of all children in the South died before reaching their fifth birthday. For children of the Civil War era, represented by this young man from the Leola area, an added obstacle for survival was about to reach every home. (Courtesy of Grant County Museum.)

# *Two*

# Legacy of the Civil War

How can fear, desperation, and a tragic ending of opportunity be accurately conveyed 150 years after the Civil War? Locally, shared memories depict an unforgiving violence; an intruder forcing from the desperate grasp of a young mother the last slab of bacon in her family's possession; Union soldiers with wagons arriving to "requisition" corn from a farmer's barn; a respected citizen and Confederate soldier, Hopkins Pratt, of Prattsville, killed by local men for a bounty for his refusal to return to war; the hiding of farm animals and family members in nearby canebrakes; the burying of valuables; and the saying of sad farewells.

Cox Creek and the Saline, swollen by heavy rain, slowed the progress of the Union retreat from Camden. April 29, 1864, saw the final deployment of an India rubber pontoon bridge placed across the Saline River at Jenkins' Ferry, which provided escape for thousands of Union soldiers at risk of being shot or captured by the Confederate army. On April 30, an awful sound of fury heard for miles destroyed hundreds of lives in the muddy fields west of the river in what was then Hot Spring County.

What destruction did the Southern stand for states' rights bring upon this land? A legacy of intimidating hate, prejudice, and a poverty of opportunity would last for a century, except for what could be created by families through their own sweaty labor.

The Civil War named this county and its seat of power. For some unreformed Confederates, honoring Northern heroes Grant and Sheridan was the final straw.

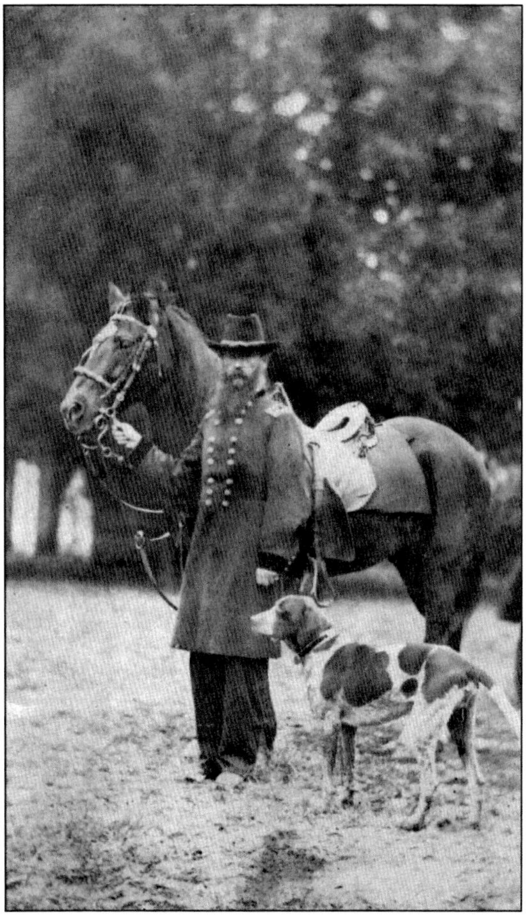

Native New Yorker and Union major general Frederick Steele retreated to Little Rock from Camden with over 10,000 soldiers just in time to catch a rain-swollen Saline River on April 29, 1864. The India rubber pontoon bridge at Jenkins' Ferry, plus efforts by Brig. Gen. Frederick Salomon's 3rd Division and Brig. Gen. John M. Thayer's Frontier Division, permitted escape from the Confederates. General Steele died in California in 1868 due to injuries sustained in a buggy accident. (Courtesy of Butler Center for Arkansas Studies.)

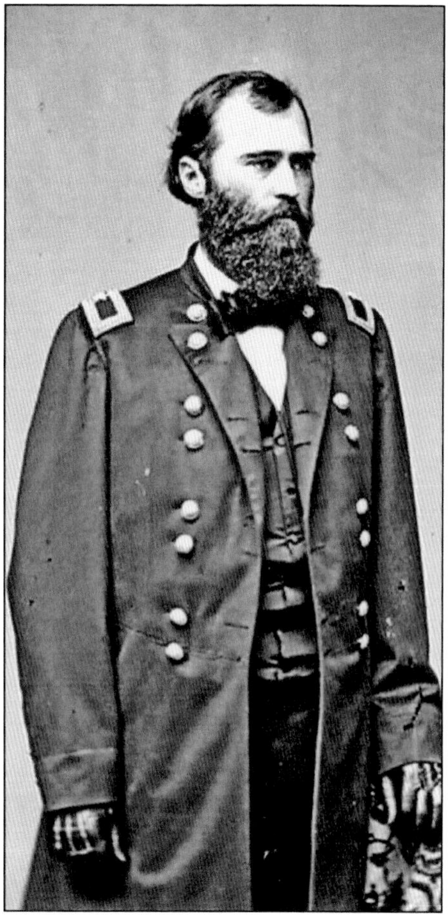

A native New Yorker, Union brigadier general Eugene Asa Carr and his cavalry were the first ordered by General Steele to cross the pontoon bridge at Jenkins' Ferry on the afternoon of April 29, 1864. His soldiers were to clear any threat on the road to Little Rock. They destroyed David Whitten's gristmill east of Prattsville. General Carr received the Medal of Honor in 1894 for three separate wounds while issuing commands at the Battle of Pea Ridge in 1862. Later in the West, he commanded military scouts Buffalo Bill Cody and Wild Bill Hickok. (Courtesy of National Archives.)

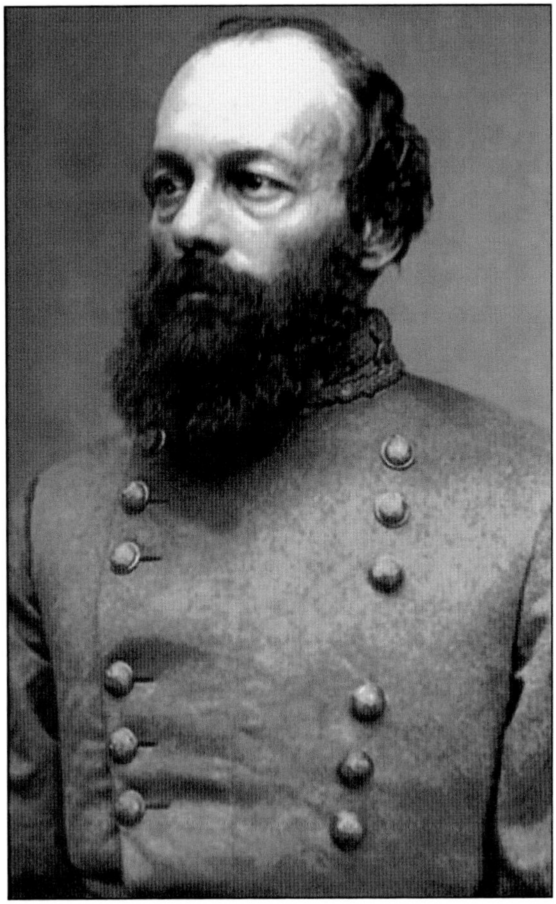

A native Floridian, Confederate general E. Kirby Smith's thousands fought the Union army along a three-mile section of the Little Rock to Camden Road between present-day Leola and Jenkins' Ferry State Park. Hundreds on both sides were killed, a few thousand wounded, yet the Federal wagons eluded forced surrender, leaving tons of supplies in the mud. After the war, General Smith was a professor of mathematics at the University of the South in Sewanee, Tennessee, and became the father of 11 children. (Courtesy of Grant County Library.)

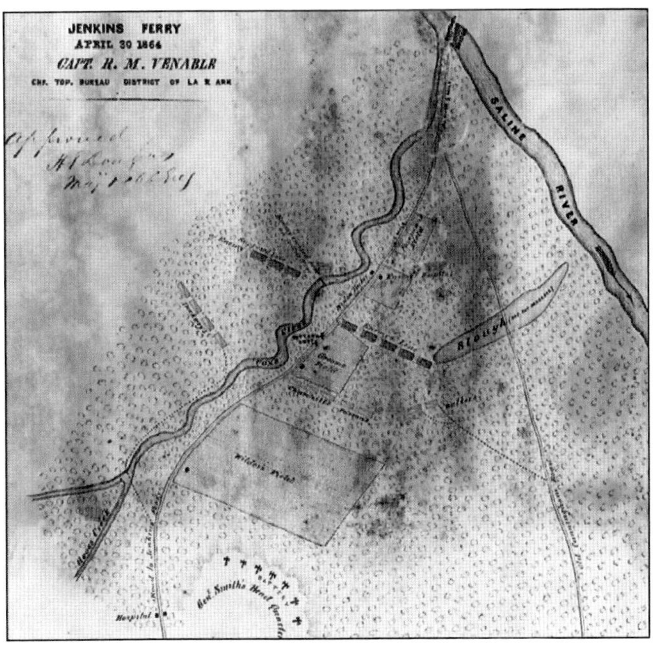


Found in the Jeremy Francis Gilmer papers at the University of North Carolina at Chapel Hill, this map of Jenkins' Ferry was drawn under the direction of Capt. R.M. Venable and sheds new light on the battle. A Groom descendant described, in May 2012 at a Saline County History and Heritage Society meeting in Benton, that a Union soldier rode up to the Groom home by their field near Cox Creek and excitedly announced that "a very big battle was about to take place right here and they should get away," so the family left and hid in a canebrake. (Courtesy of Dr. Bill J. Gurley.)

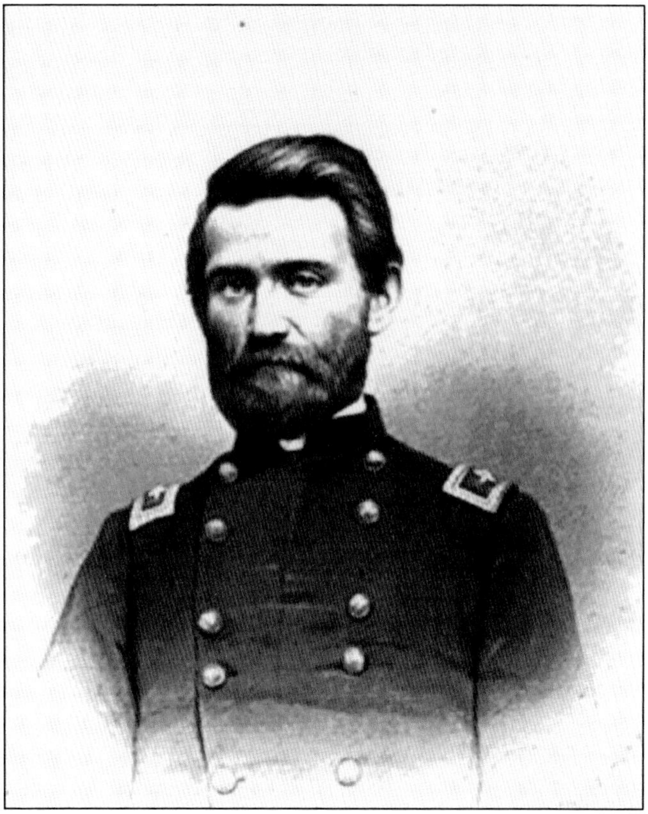

A native New Yorker, Union brigadier general Samuel Allen Rice had served as only the second attorney general of Iowa from 1856 to 1861. While leading soldiers in Brigadier General Salomon's brigade in the last major skirmish at Jenkins' Ferry, General Rice was shot in the right ankle, resulting in metal from his spur being shot deep into his foot. Traveling in the ambulance train headed to Pine Bluff through present-day Crossroads on April 30, 1864, he was able to reach home in Oskaloosa, Iowa, on May 18 but died on July 6. (Courtesy of Mahaska County Historical Society, Oskaloosa, Iowa.)

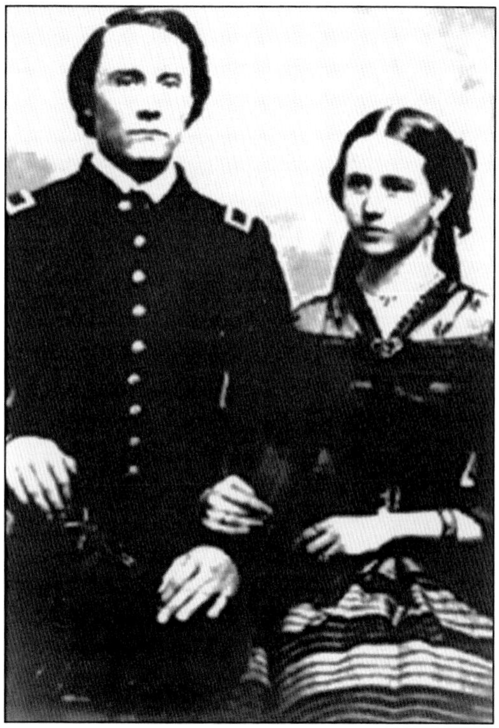

Confederate general Horace Randal, a graduate of West Point, is pictured with his first wife, Julia S. Bassett, whom he married on June 2, 1858, in New London, Connecticut. Julia died in 1860 from the rigors of military camp life in New Mexico. General Randal personally recruited the 28th Texas Cavalry Regiment at Marshall, Texas. On July 8, 1862, he married Nancy Taylor of Marshall. After his death at Jenkins' Ferry, the county surrounding Amarillo, Texas, was named in his honor. (Courtesy of Harrison County Historical Museum, Marshall, Texas.)

Confederate brigadier general William Read Scurry, member of the Congress of the Republic of Texas, Mexican War veteran, and Austin newspaper owner, invaded New Mexico during the Civil War, recaptured Galveston, and fought at the Battle of Mansfield in Louisiana. After being wounded at Jenkins' Ferry, he asked if the Yankees were whipped. Told that they were, he replied, "Now take me to a house where I can be made comfortable and die easy." Texas named a county southeast of Lubbock in his honor. General Scurry and wife, Janette Sutton Scurry, were the parents of seven. (Courtesy of Snyder County Museum, Snyder, Texas.)

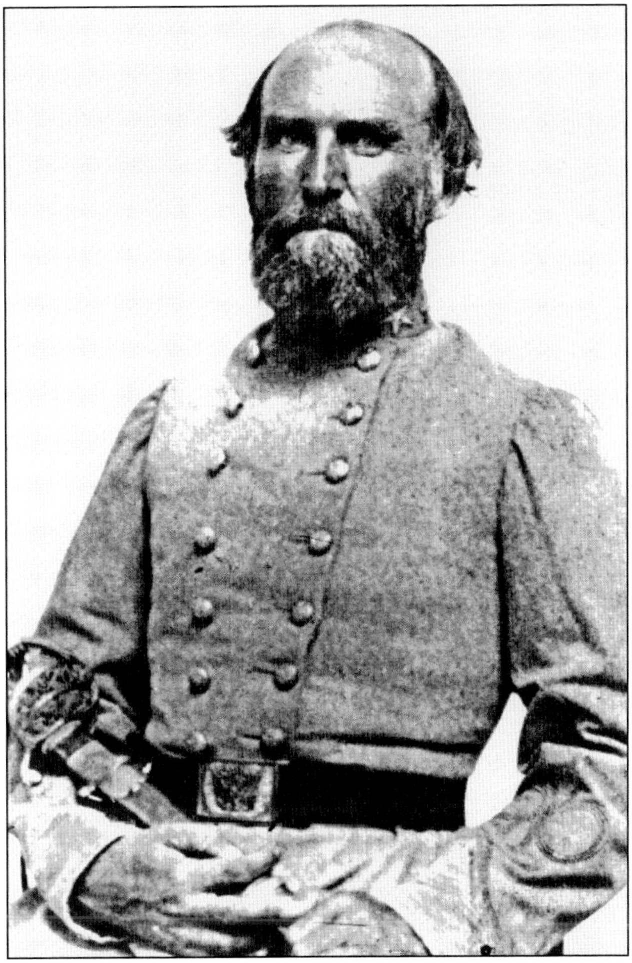

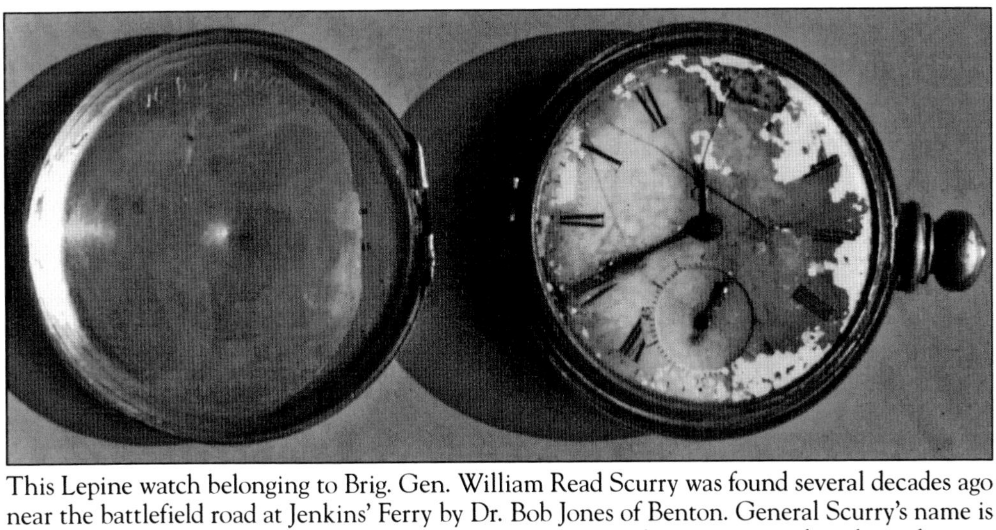

This Lepine watch belonging to Brig. Gen. William Read Scurry was found several decades ago near the battlefield road at Jenkins' Ferry by Dr. Bob Jones of Benton. General Scurry's name is near the top of the inside cover. It was found under leaves after a century of marking the spot where Scurry was wounded. (Courtesy of Dr. Bill J. Gurley.)

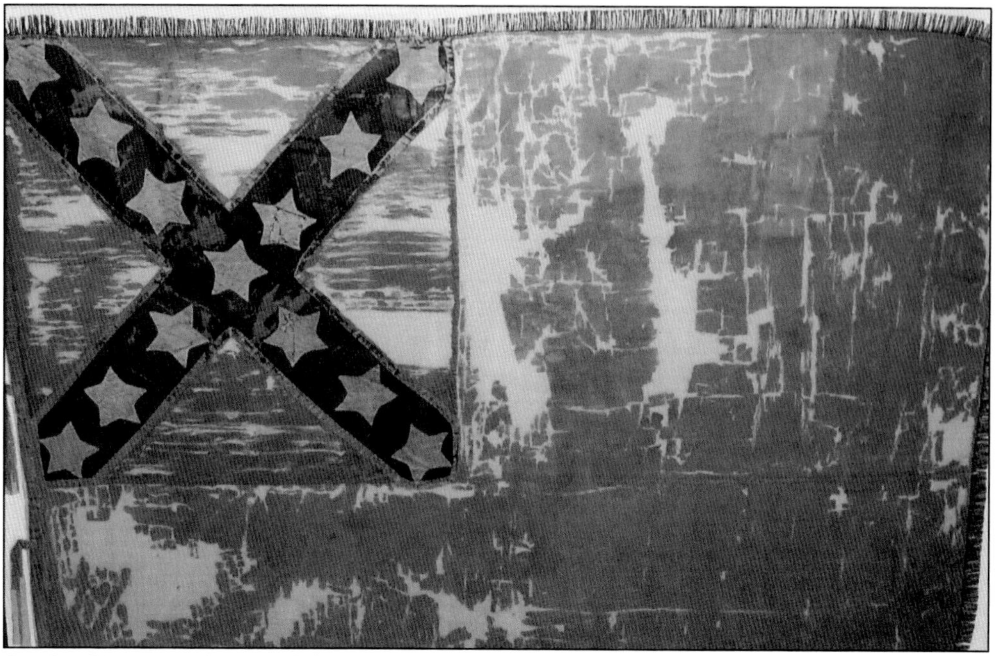

On April 30, 1864, this Confederate flag was taken by wagoner John Welhaupt and Pvt. William Ohler, both of Company B, 9th Wisconsin Regiment. Presented to Wisconsin governor James Taylor Lewis by General Salomon, it was "deposited with other trophies in the state capital." In the first decade of the 21st century, the flag was presented to the Arkansas National Guard Museum at Camp Robinson, Little Rock. Two other flags were captured at Jenkins' Ferry by the 50th Indiana. (Courtesy of the Army National Guard Museum.)

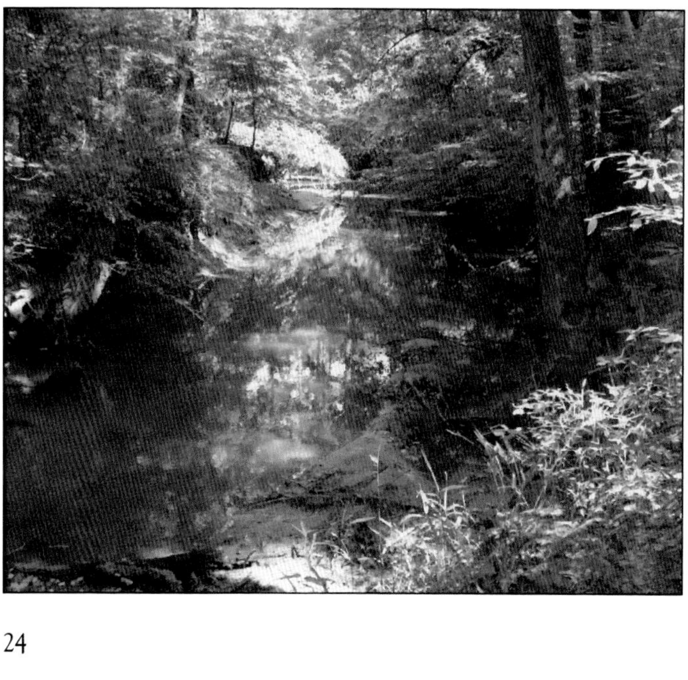

Cox Creek, then part of Hot Spring County, was a center of fighting during the Battle of Jenkins' Ferry on April 30, 1864. The Little Rock to Camden Stagecoach Road, which ran just south of the creek for three miles, hosted intense fighting west of the Saline River. Both Union and Confederate units crossed the creek to fight near its north bank. This photograph was taken on April 29, 2012—the 148th anniversary of the battle's opening shots. (Courtesy of the author.)

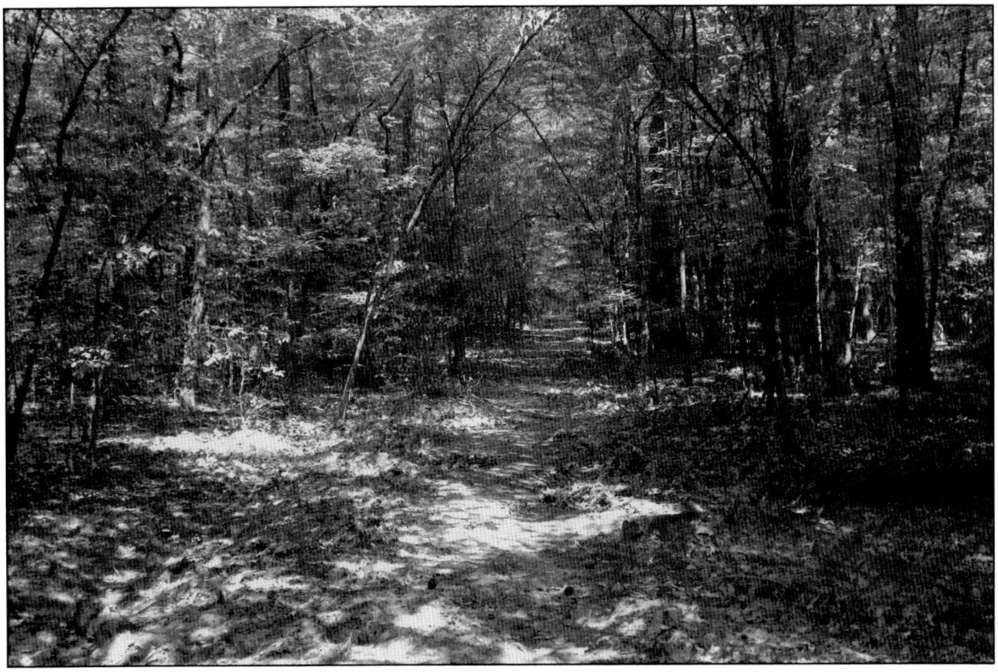

This section of the Little Rock to Camden Stagecoach Road, built in the 1840s, runs just south of Cox Creek and was the scene of unimagined horror as Union soldiers struggled with the Confederate attack. Union soldiers had to deal with hundreds of refugees and sutlers, then tend to hundreds of wounded, of which a number were abandoned on the battlefield. Tons of supplies, hundreds of six-mule wagons, and additional horses and mules were lost. (Courtesy of the author.)

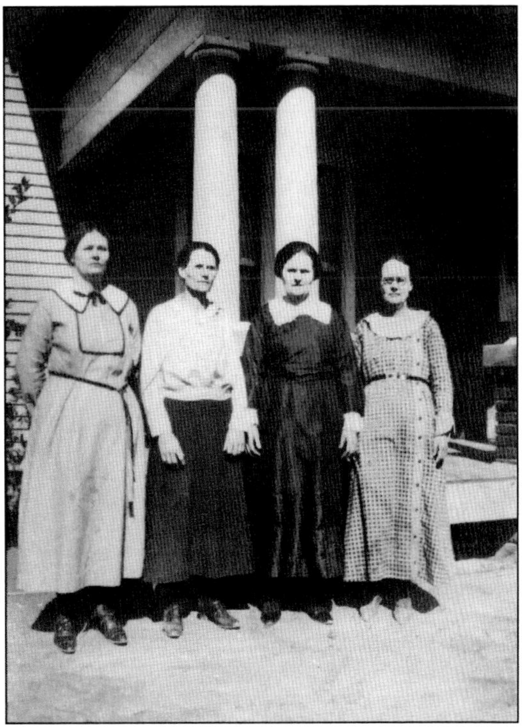

Tennessee native William R. Barnes left Cherry Grove in southern Saline County to join Company K, 2nd Arkansas Infantry Regiment, CSA. Wounded above the right knee at the Battle of Shiloh, he was captured and admitted to the USA Hospital at Camp Dennison, Ohio, on April 18, 1862. He reported "having many of his buddies shot within one foot of him" and eating "bessie bugs and elderberries to keep from starving." He returned home, marrying Matilda Jane Bratton on June 25, 1867. Shown here are four of their children, from left to right, Nina, Cora, Melissa, and Ida. (Courtesy of the author.)

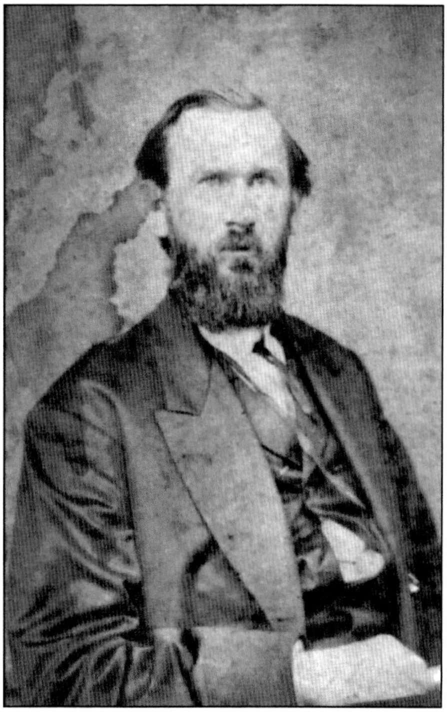

Dr. Richard Clinton Rhodes and his wife, Susan, of Clinton, North Carolina, settled in southern Saline County in the 1850s. Named after his maternal grandfather, Revolutionary War veteran Lt. Col. Richard Clinton, Rhodes had inherited a large estate from his father, Jacob, in North Carolina. Dr. Rhodes was hanged by a group demanding money; he survived but died of complications months later in 1867. (Courtesy of Grant County Museum.)

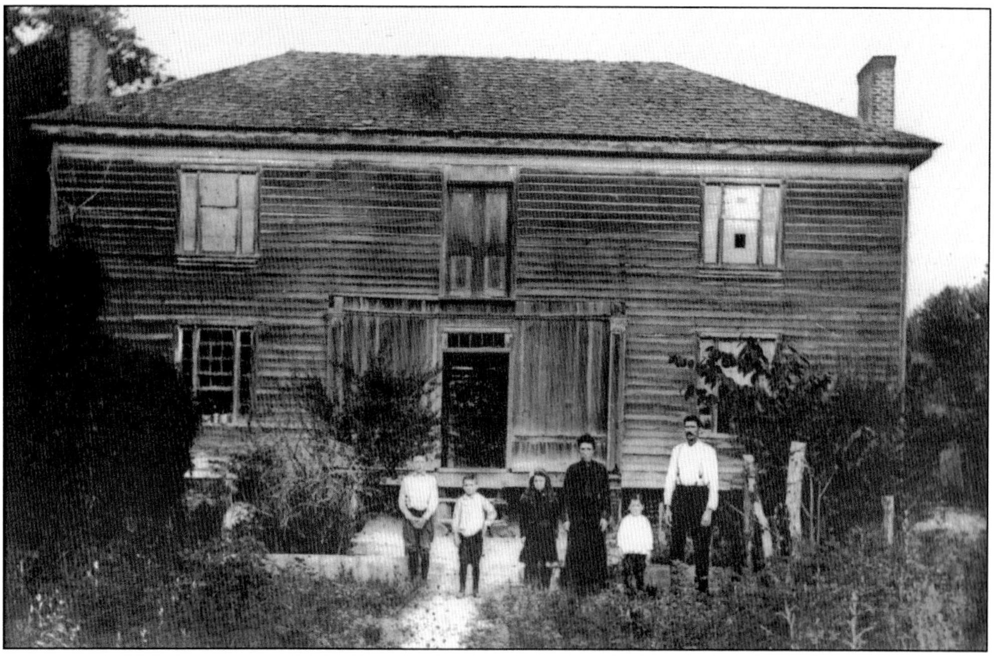

Far larger than any area dwelling, the home of Dr. Richard C. Rhodes, located eight miles south of present-day Sheridan, reportedly served wounded Union and Confederate soldiers after the Battle of Jenkins' Ferry. The plantation home, site of a Confederate barbecue during the Civil War, had a detached kitchen with walkway to the north. A large gazebo and slave quarters were behind the house, and a barn was to the south. Pictured around 1915 are family members, from left to right, Busby, Sam, Lula, Sarah, Easter, and Jesse Rhodes. (Courtesy of Grant County Museum.)

# *Three*

# Reconstruction's Gift

In 1868, rumblings concerning a new county became more than just talk. Citizens near the Saline River dreaded long wagon journeys to visit courthouses in Benton, Malvern, or Pine Bluff. Legislation was drawn up to "form a separate and distinct county, to be named Grant. . . . That the temporary seat of justice shall be at the school-house situated on the northeast quarter of the northwest quarter of section two, township five south, range 13 west, in Merry Green township [southwest of the present junction of Little Creek Cutoff and State Highway 46]. That Thomas A. Morris and William M. Allison, of Saline county, and J.W. Harrison, of Hot Spring county, shall constitute a board of commissioners for the county of Grant, whose duty it shall be to locate the permanent seat of justice . . . not to exceed five miles from the center thereof. . . . That when permanently located, the said county seat shall be called Sheridan. Approved February 4, 1869." One month later to the day, former general Ulysses S. Grant became president of the United States.

Commissioner William Martin Allison, a native of Missouri, served with Union general Steele at Pea Ridge and Jenkins' Ferry, married the daughter of E.H. Vance Sr., and practiced medicine at Bee Branch, Quitman, and Conway. Saline County judge and commissioner Thomas A. Morris became the first Grant County judge, and Commissioner Dr. John W. Harrison continued practicing medicine.

Littleton Meek Veazey donated 40 acres in 1869 to establish that "seat of justice" in the new town of Sheridan. On land purchased in 1857, his farm was described in 1889 as "one hundred cleared acres, large buildings, and an orchard," which became the eastern side of the city.

On June 28, 1873, in Little Rock's *Daily Gazette*, W.N.C. from Grant County reported: "Grant. What a name to inspire emotions of gratitude in the breast of a tax-ridden people, and our county seat, Sheridan, another military accident, who has immortalized himself by his infamous career in the Shenandoah Valley during the late war, whose very name awakens reminiscences of the days of Nero and Caligula." A later contributor accepted the Union names due to "Gen. Grant's liberal terms of surrender and kindness shown to Gen. Lee."

In November 1865, Elizabeth White, mother of Jane Pinchback, was tortured to death south of the Saline by local men desperate for money. Months later, south of Crossroads, Dr. Richard Clinton Rhodes died from wounds sustained in a hanging by local men with similar reasons. In December 1873, David James Wilson was murdered by a nephew by marriage at the Ten Mile Wagon Yard at Camp Creek Branch while returning from Pine Bluff. In 1880, Sheridan's population was 42, growing to 184 in 1890. The City of Sheridan was incorporated on August 26, 1887, with official papers signed by William Thomas Veazey, son of the man who donated land for the town. A night watchman was soon hired.

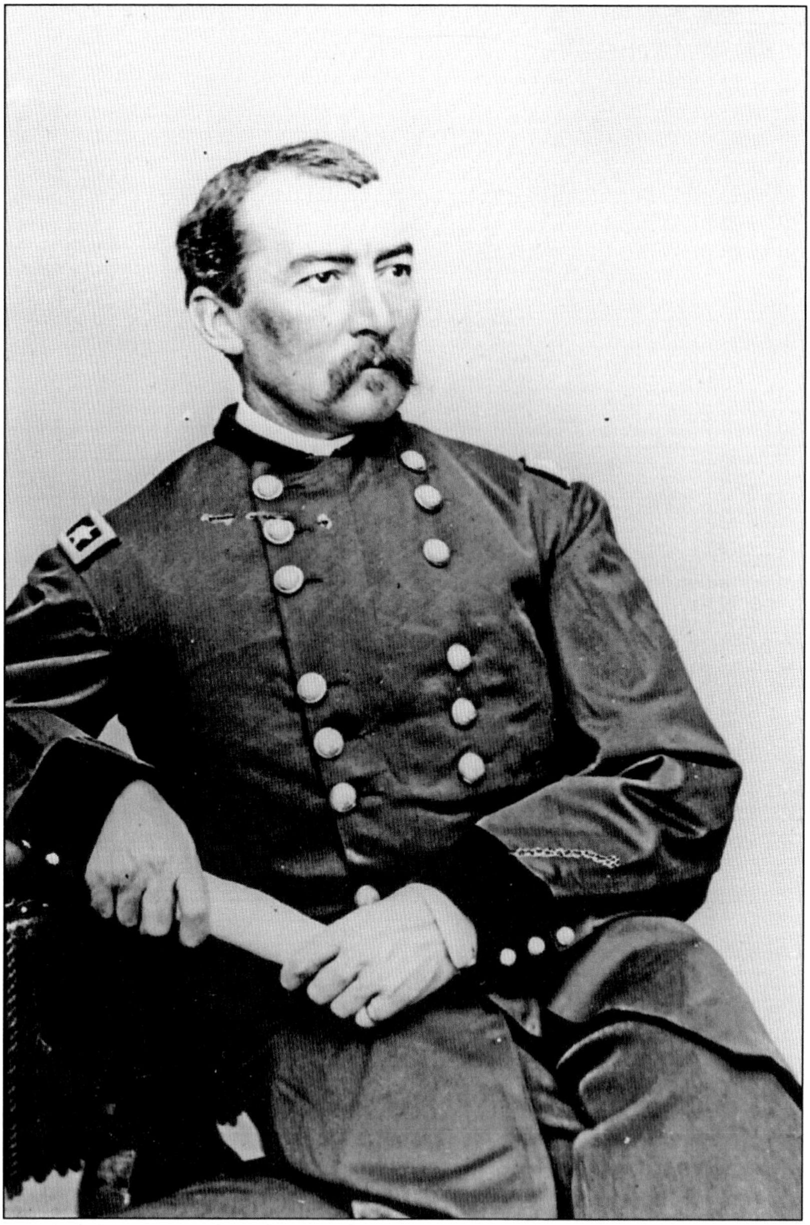

Union general Philip Henry Sheridan was born in New York of Irish immigrant parents in 1831. Nicknamed "Little Phil" due to his height (five feet, five inches), he graduated from West Point in 1853, thirty-fourth in a class of 52, after being suspended a year for fighting. During the Civil War, General Sheridan was known for his "scorched earth" policy and "Sheridan's Ride" in Shenandoah Valley, Virginia. In 1870, President Grant sent him to observe the Franco-Prussian War, and he saw Napoleon III surrender. In 1871, he was in Chicago during the Great Fire and led military relief efforts. Sheridan's personal crusade was to protect Yellowstone's natural wonders, once ordering soldiers to take control. He conducted the Indian Wars in the West in the 1870s. Counties were named in his honor in North Dakota, Nebraska, Montana, Wyoming, and Kansas, and cities in Colorado, Montana, Wyoming, Oregon, Indiana, and Illinois, as well as Sheridan, Arkansas. (Courtesy of National Archives.)

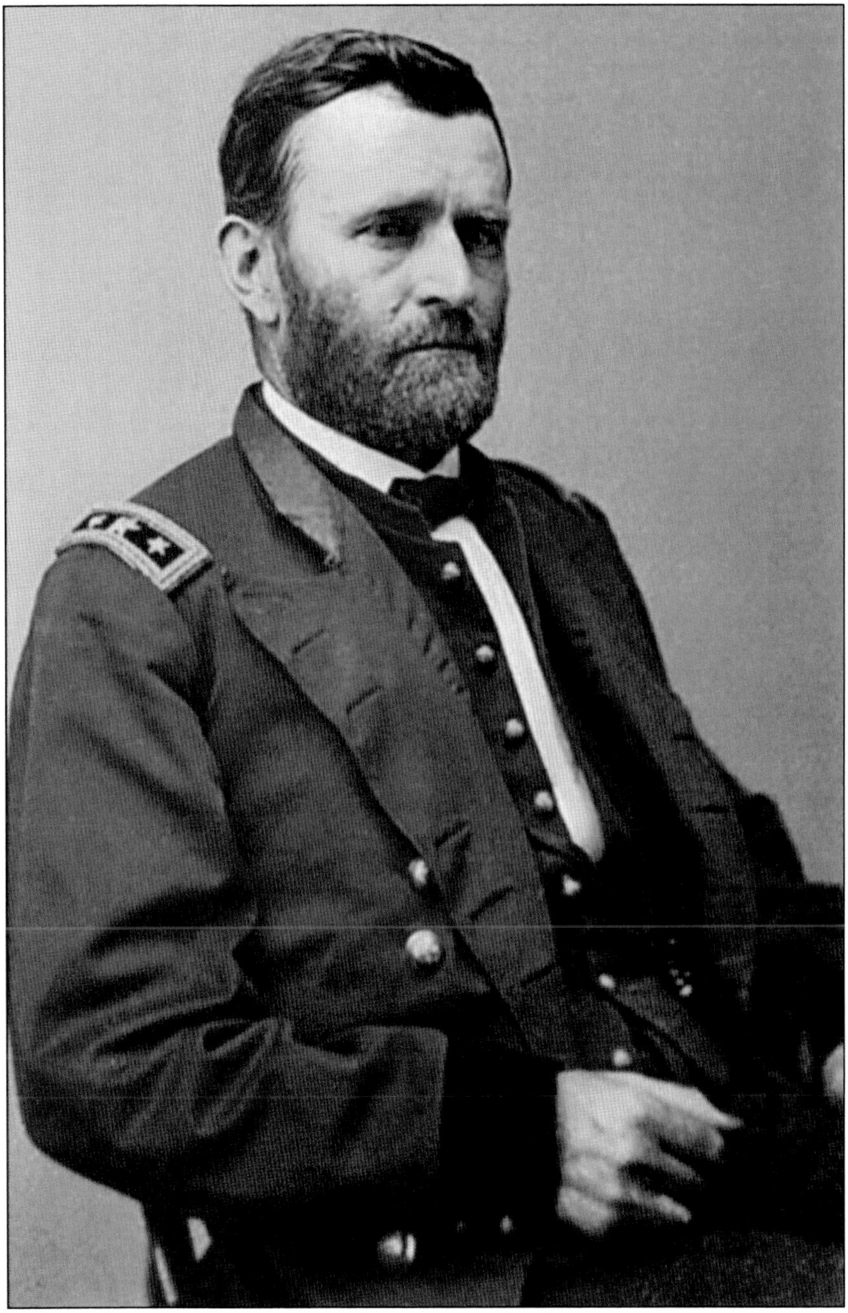

Union general Ulysses S. Grant, born in Ohio in 1822 of English and Scottish ancestry, graduated from West Point in 1843, ranked 21st out of 39. He was a Mexican War veteran and Civil War Union commander. In 1868, he was elected president of the United States just as legislation creating Grant County was being readied at the Union-controlled Old State House in Little Rock. After serving as president, Grant and wife Julia visited Little Rock on April 15, 1880, spoke after a military review, and stayed at the Capital Hotel. Counties and cities were named for Grant in Kansas, Wisconsin, Indiana, and Alabama, as well as Grant County, Arkansas. (Courtesy of National Archives.)

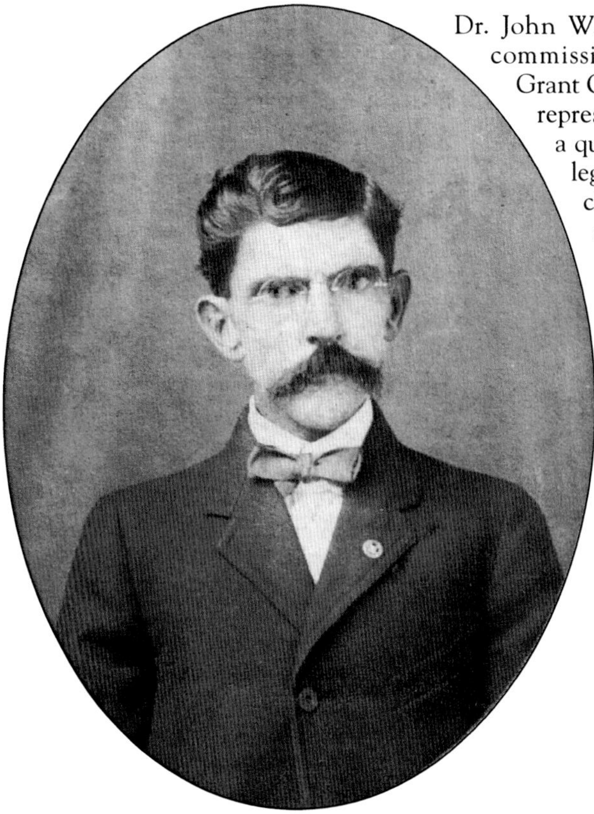

Dr. John W. Harrison served as one of three commissioners named in the law creating Grant County on February 4, 1869. As a state representative during 1868, Dr. Harrison, a quiet Union sympathizer, helped move legislation along, partly explaining the choice of honoring Union generals Grant and Sheridan. He practiced medicine in Leola, a community he knew as Sandy Springs. (Courtesy of Grant County Museum.)

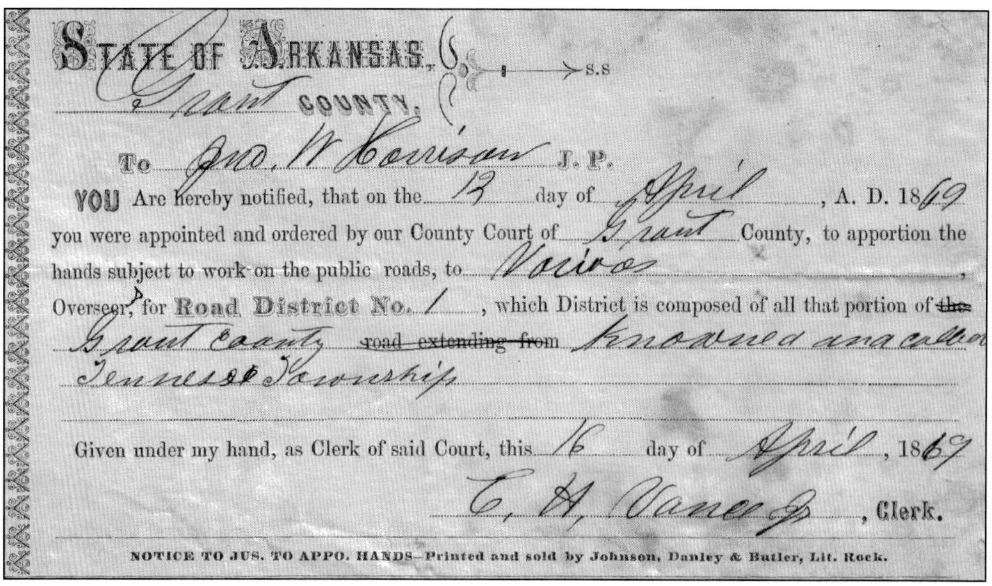

STATE OF ARKANSAS, Grant COUNTY, } s.s

To Jno. W Harrison J. P.

YOU Are hereby notified, that on the 12 day of April, A. D. 1869 you were appointed and ordered by our County Court of Grant County, to apportion the hands subject to work on the public roads, to Various, Overseers for Road District No. 1, which District is composed of all that portion of ~~the~~ Grant County ~~road extending from~~ known and called Tennessee Township.

Given under my hand, as Clerk of said Court, this 16 day of April, 1869

E. H. Vance Jr, Clerk.

NOTICE TO JUS. TO APPO. HANDS—Printed and sold by Johnson, Danley & Butler, Lit. Rock.

This April 16, 1869, document appointed Dr. John W. Harrison of Leola to be in charge of managing road crews in Tennessee Township. In lieu of paying taxes, men could donate three days of labor each year to the county, helping improve roads, a practice that continued until the 1940s. It was signed by the first circuit and county clerk for Grant County, E.H. Vance Jr. (Courtesy of Grant County Museum.)

Thomas A. Morris purchased 85 acres in section 5, township 6 south, range 13 west on July 1, 1861. The land was located west of Crossroads, in what was then southern Saline County. He was serving as Saline county judge when appointed one of three commissioners to oversee the creation of Grant County. Morris was elected the new county's first judge. (Courtesy of Grant County Museum.)

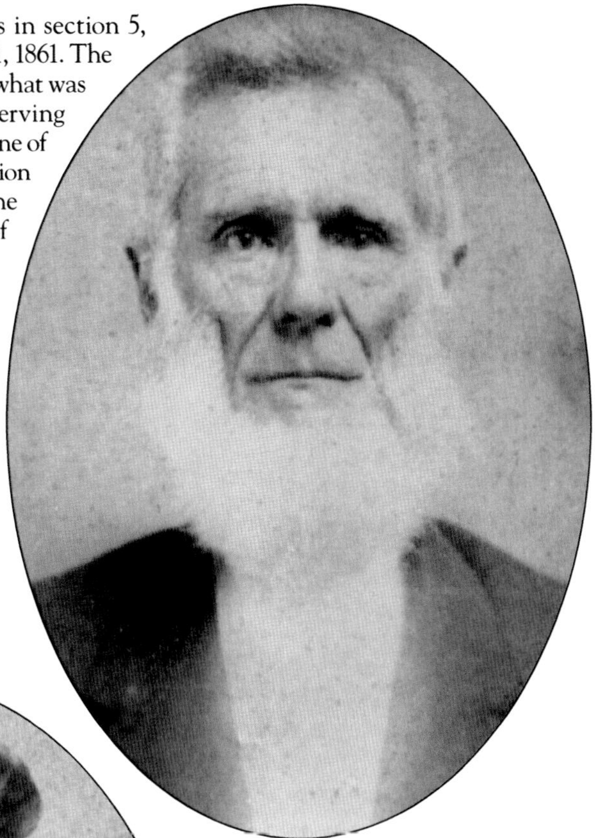

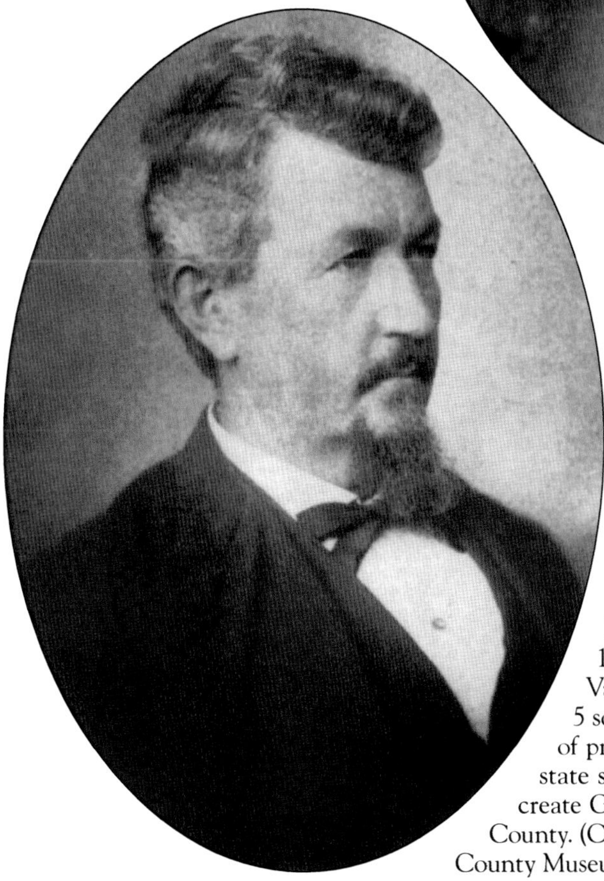

Enoch Hugh Vance Jr., a Mississippi native, attended Abington College in Illinois before becoming Saline County assessor, first circuit, county and probate clerk of Grant County, and county judge. He then served as lawyer and postmaster in Malvern. In 1859 and 1860, his father, Enoch Harvey Vance, purchased 320 acres in township 5 south, range 13 west, located southeast of present-day Sheridan. Enoch Sr. was a state senator from Saline County, helping create Grant County before moving to Perry County. (Courtesy of Patti Vance Hays and Grant County Museum.)

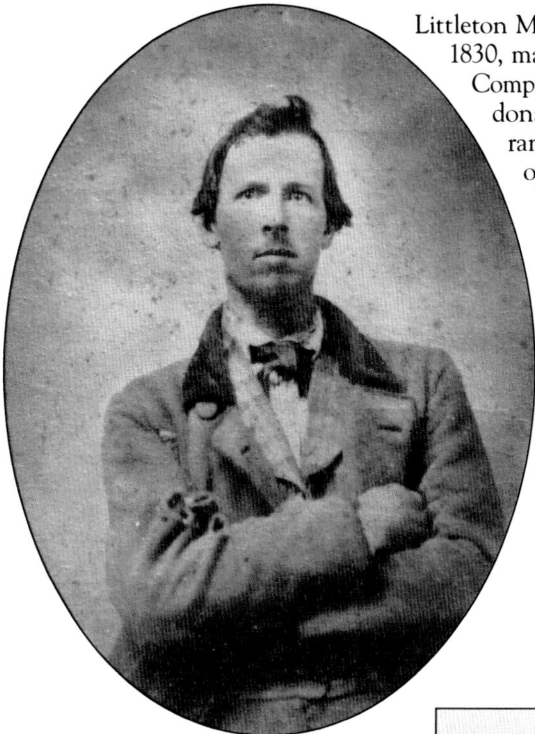

Littleton Meek Veazey, born in Alabama on March 9, 1830, married Mary E. Clark in 1856 and served in Company H, 18th Arkansas Infantry, CSA. He later donated land in section 10, township 5 south, range 13 west for construction of Sheridan, about one mile southwest of the schoolhouse where legislation mandated the first county meeting be held to organize Grant County. (Courtesy of Grant County Museum.)

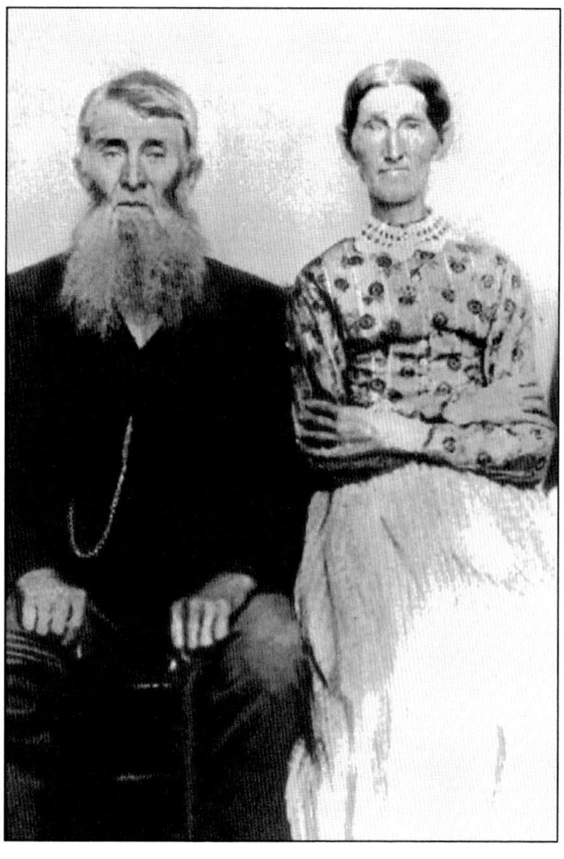

Nancy Jane Veazey (right) was born on July 8, 1833, in Alabama to John Veazey and Nancy King. She married John Hamilton (left) on April 28, 1850, and they became the parents of eight. Nancy Jane was the youngest sister of Littleton Veazey, who donated part of his farm to create the city of Sheridan. (Courtesy of Benny Barnes.)

Seaborn D. Reese, born in 1821 in Wilkerson County, Georgia, purchased 320 acres on July 2, 1860, in section 14, township 4 south, range 13 west, located four miles north of Sheridan. During the Civil War, he served in Company A, 11th Arkansas Infantry, CSA, commanded by Capt. M. Vance. He was Grant County's second sheriff and died on September 14, 1895, while mayor of Sheridan. He was survived by his wife, Celia, and seven of nine children. (Courtesy of Grant County Museum.)

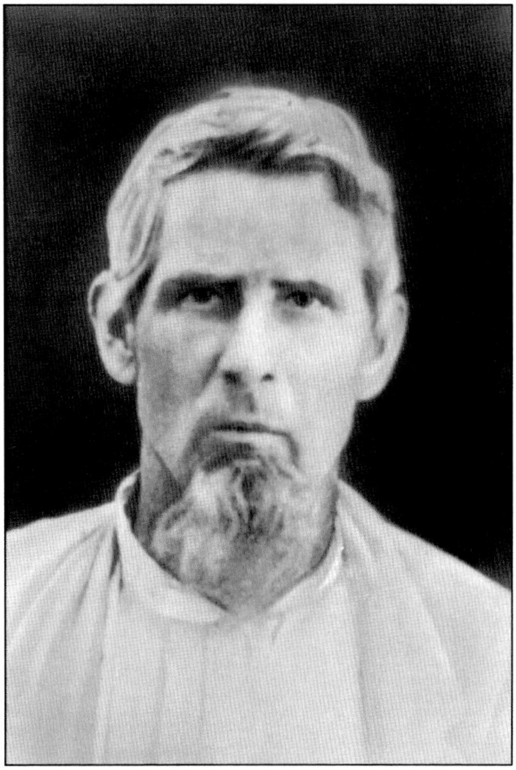

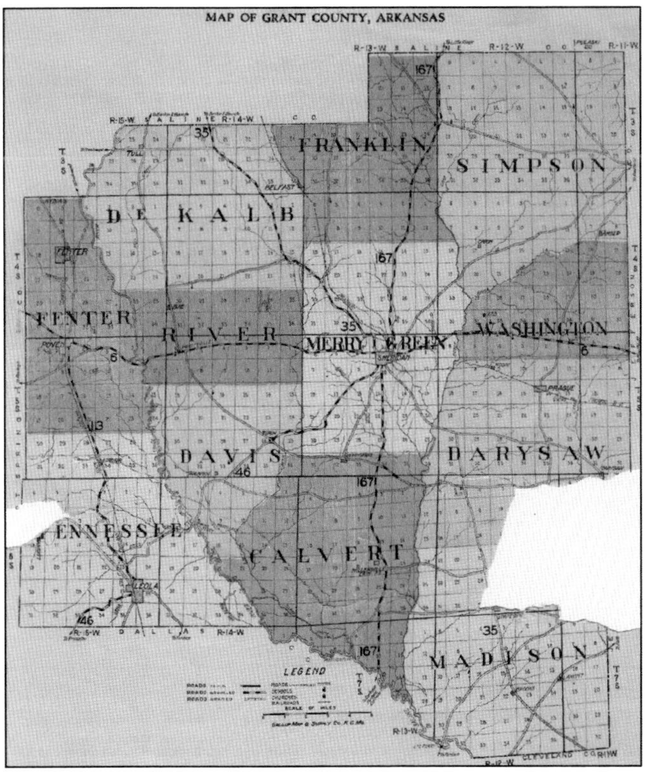


Published a century ago, this township map of Grant County was sold by the Gallup Map and Supply Company of Kansas City, Missouri. It shows the old Little Rock to Camden Stagecoach Road still in use, south of Cox Creek past Jenkins' Ferry. Timber communities such as Brooks, Lamont, Millersville, Arthur, Nydia, Koons, Amber, Ebb, and Buie disappeared when sawmills closed. (Courtesy of the author.)

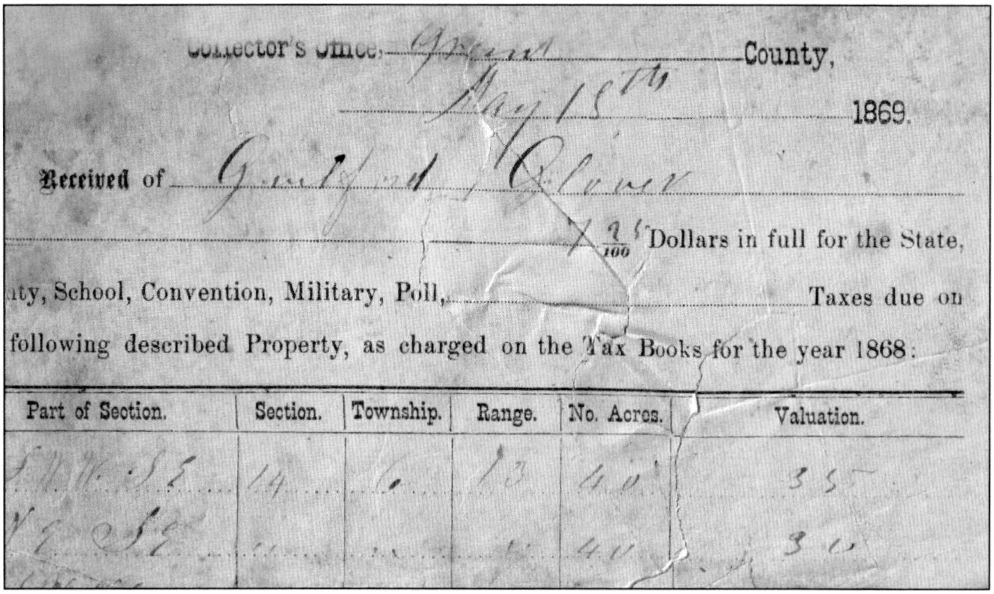

Collector's Office, Grant County,

May 18th 1869.

Received of Guilford Glover

$7\frac{25}{100}$ Dollars in full for the State,

...ty, School, Convention, Military, Poll, Taxes due on

following described Property, as charged on the Tax Books for the year 1868:

| Part of Section. | Section. | Township. | Range. | No. Acres. | Valuation. |
|---|---|---|---|---|---|
| [illegible] | 14 | 6 | 13 | 40 | 35 |
| [illegible] | " | " | " | 40 | 30 |

One of the earliest tax receipts in existence from newly created Grant County is this May 18, 1869, document showing that Guilford Glover paid $7.25 due on over 120 acres in section 14, township 6 south, range 13 west, located two miles south of Crossroads between Hurricane Creek and Highway 167. (Courtesy of Lessie Wallace.)

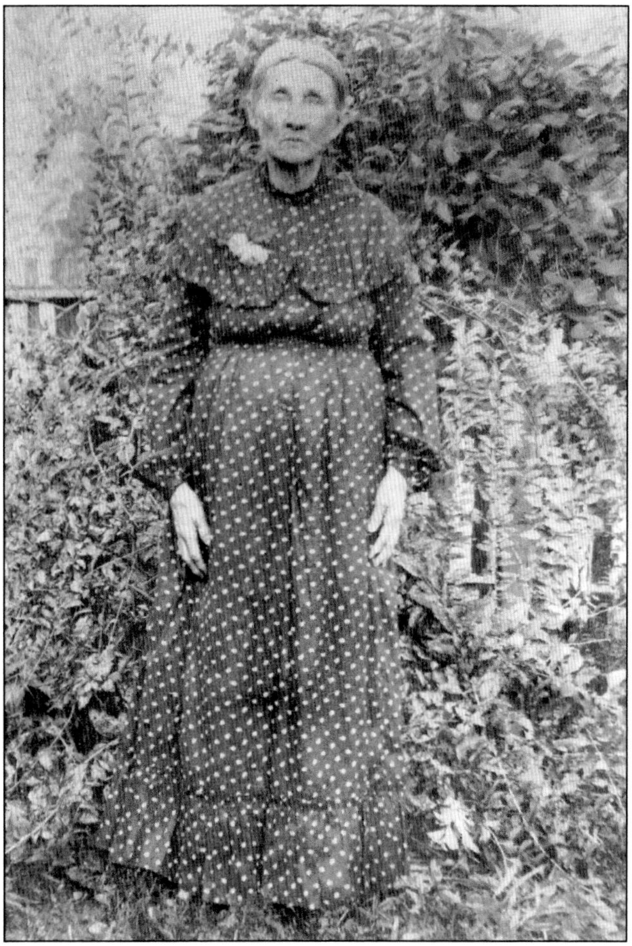

Betheny Morgan Wilson, a native of Alabama, married David James Wilson in April 1850 in Lowndes County. In the Civil War, he was a member of the 53rd Alabama Partisan Rangers. The Thursday, December 18, 1873, edition of the *Pine Bluff Press* reported: "A man named D.J. Wilson was killed by his nephew on Tuesday night while on his way from this city to his home in Grant County." Betheny was left alone with small children during a troubled time. (Courtesy of the author.)

Samuel Wright Allen and his sister, Rebecca Allen Jones, pose at her son's house at the southwest corner of South Main and Holly Streets in Sheridan. Both survived the Civil War, with Sam being forced by the "bluecoats" in 1864 to rake corn from his father's barn into Union army wagons at their parents' first Arkansas home south of the Saline River. Rebecca's son and daughter-in-law, Lem and Claudia Reese Jones, ran a successful store on Center Street. (Courtesy of the author.)

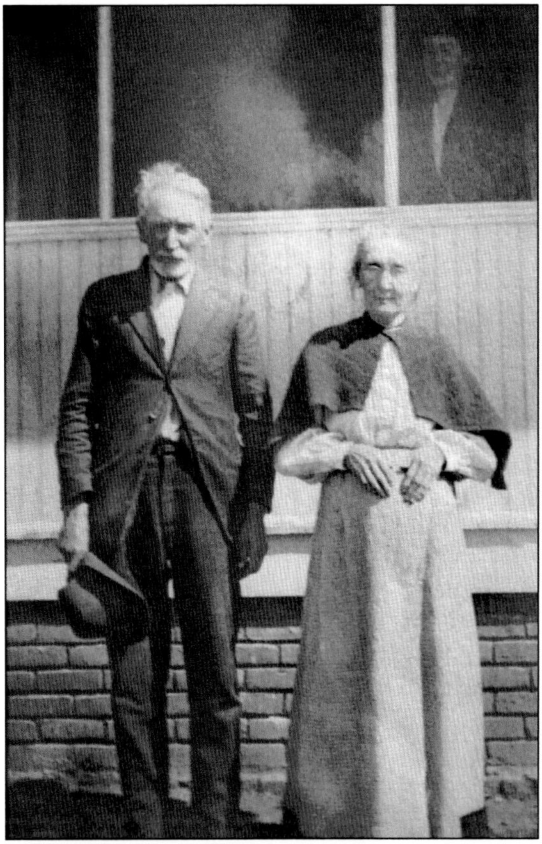

Two children enjoy an early-20th-century adventure with their parents in this photograph from the papers of Dr. John W. Harrison of Leola. (Courtesy of Grant County Museum.)

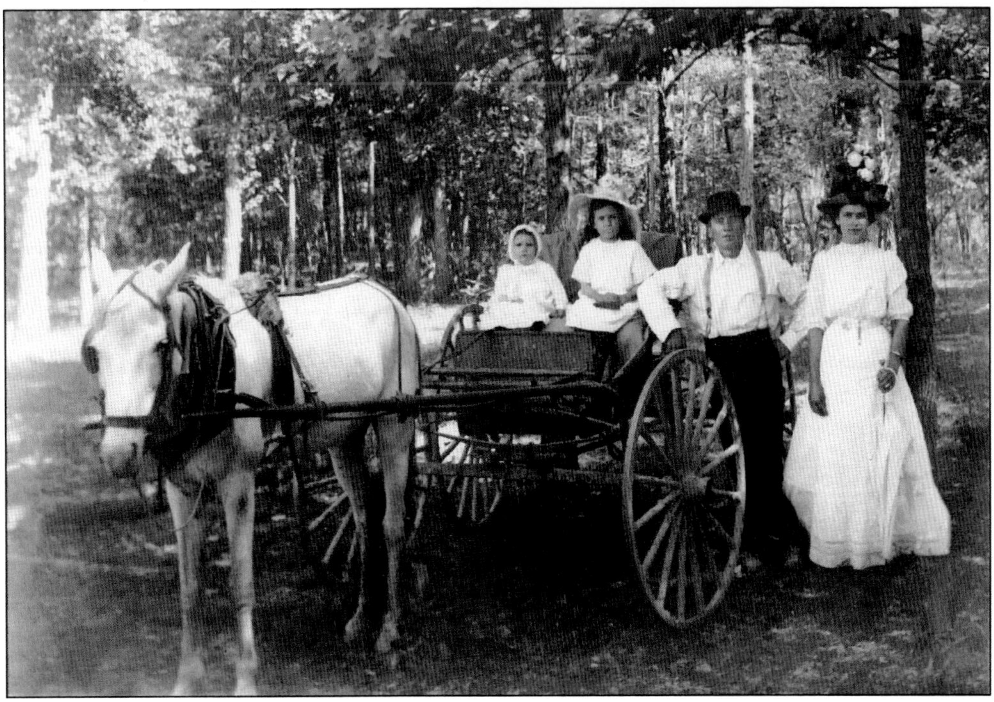

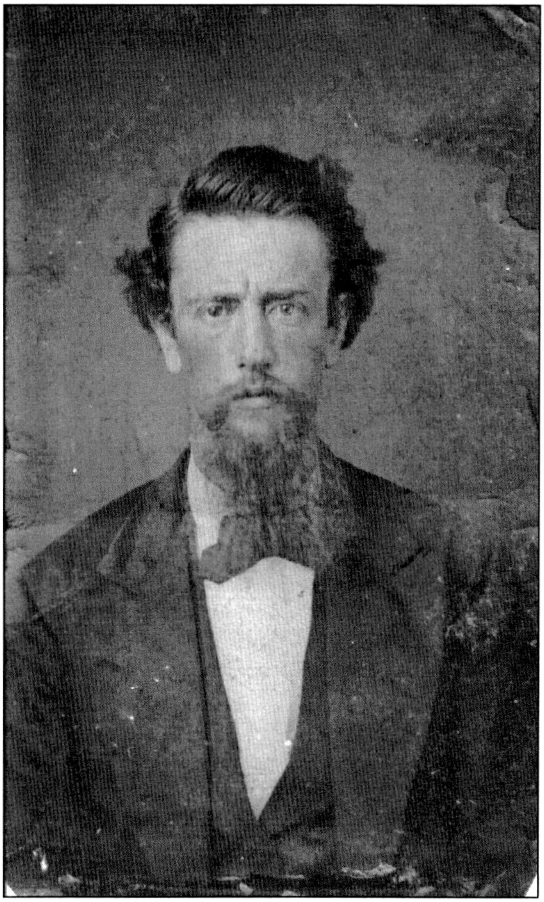

Among the papers of Dr. John W. Harrison is this photograph from the 1800s of Daniel Samuel Harrison of Leola, then known as Sandy Springs. (Courtesy of Grant County Museum.)

At the corner of Barraque and Bond Streets in Pine Bluff, Taylor & Cleveland sold essentials such as these in 1876: coffee (at 22 1/2¢ per pound), rice (at 8¢ a pound), and pork (at 10¢ a pound). Among the purchases by P.R. Rhodes shown here were powder, shot, and muzzleloader caps. The total bill was $58.30. (Courtesy of Grant County Museum.)

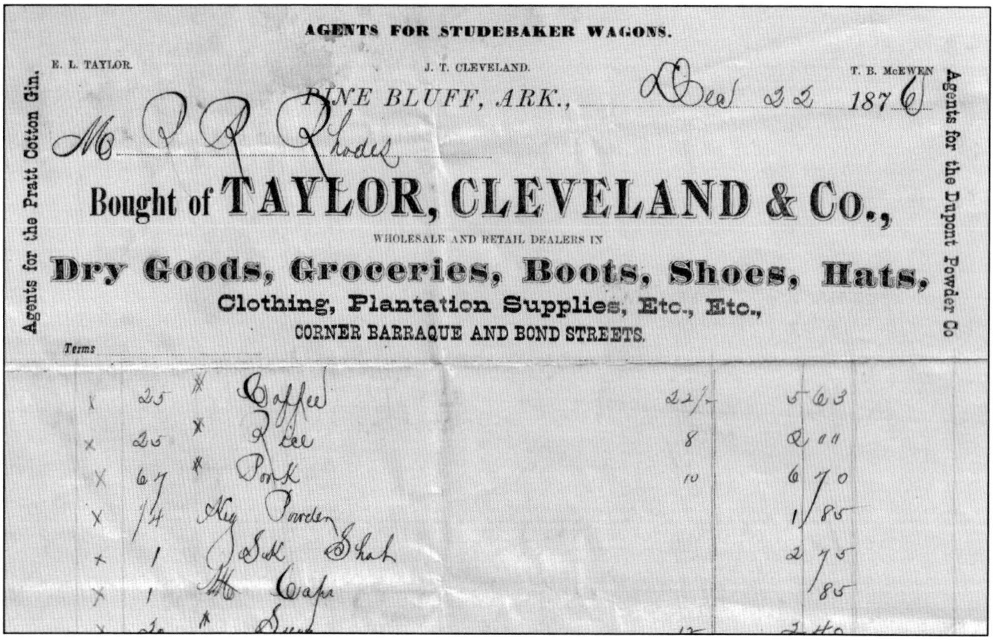

AGENTS FOR STUDEBAKER WAGONS.

E. L. TAYLOR. J. T. CLEVELAND. T. B. McEWEN

Agents for the Pratt Cotton Gin.

PINE BLUFF, ARK., Dec 22 1876

M P R Rhodes

Bought of TAYLOR, CLEVELAND & Co.,

WHOLESALE AND RETAIL DEALERS IN

Dry Goods, Groceries, Boots, Shoes, Hats,

Clothing, Plantation Supplies, Etc., Etc.,

CORNER BARRAQUE AND BOND STREETS.

Agents for the Dupont Powder Co

Terms

| | Qty | Item | Price | Amount |
|---|---|---|---|---|
| x | 25 | Coffee | 22 1/2 | 5 63 |
| x | 25 | Rice | 8 | 2 00 |
| x | 67 | Pork | 10 | 6 70 |
| x | 1/4 | Keg Powder | | 1 85 |
| x | 1 | Sack Shot | | 2 75 |
| x | 1 | M Caps | | 85 |

# *Four*

# Education's Progress

In the 1870s, Reconstruction-era Arkansas finally began a statewide public school system, which gradually created thousands of one-room schoolhouses. Almost 50 would exist in Grant County. During those years, a family's annual school tax might be 12¢. For several decades, if students wanted to progress with their education, they would have to attend high school outside the county or live in Sheridan, either attending the town school or taking additional courses at the Missionary Baptist College while it was in operation from 1921 to 1934. Teachers were required to pass a test and attend teacher institutes to receive a license. A county school superintendent maintained an office at the courthouse.

The methods in these one- or two-room schools, housing all eight grades, were effective through repetition. Lessons were checked at a recitation bench, where each grade of students waited its turn to recite for the teacher. School terms of three months' duration were scheduled throughout the year to match seasons of farmwork, because students were needed at home. School was viewed as a vacation from hard farm labor and as a way up and out for those who were supported in their efforts.

Consolidation of smaller districts with Sheridan began with Lost Creek, Marlow, and Liberty (in 1928), followed by Bois d'Arc, Crossroads, and Stuckey (1929), Corinth (1930), Wilson (1931), Hood (1935), Little Creek (1936), Moore (1939), Palestine, Cane Creek, and Shiloh (1940), Sandy Ridge and Tabor (1941), Carder (1946), Orion (1947), and Oak Grove, Walnut Ridge, and East End (1949). School districts at Tull, Shaw, and Mount Olive were added to Bauxite (1949). Also joining Sheridan were Grapevine (1986), Leola (1987), and Prattsville (1994). Whether attained formally or informally, education is how people better their odds at having a future of unimagined dreams. Most of the residents of Sheridan and Grant County know this and have worked diligently to lift themselves up to a better standard of living. That is the story of this town and county, how its people made a lot out of a little, proving that education's progress is their chance at a better life.

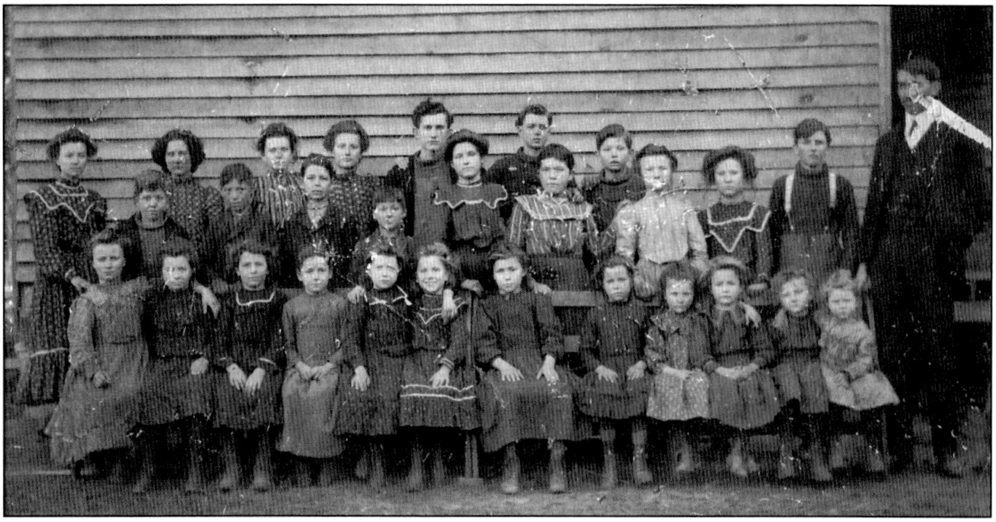

Wilson, or 16 School, located west of early-1900s Crossroads resident Bradberry B. Wilson's home, had a wooden block foundation with a space measuring 20 by 30 feet for the primary grades and a 20-square-foot room on the northwest for older students. The building was furnished with a blackboard and a recitation bench. Buddy Ault attended all eight grades, remembering teachers Bill McClellan, Pat Crouse, Charlie Mathis, Elvie Johnson Crouse, Clinton Lovell, Bea Black, Mancil Wilson, Daisy Camp, Ava Ashcraft, Gus McMillan, and Ed McCool. (Courtesy of Grant County Museum.)

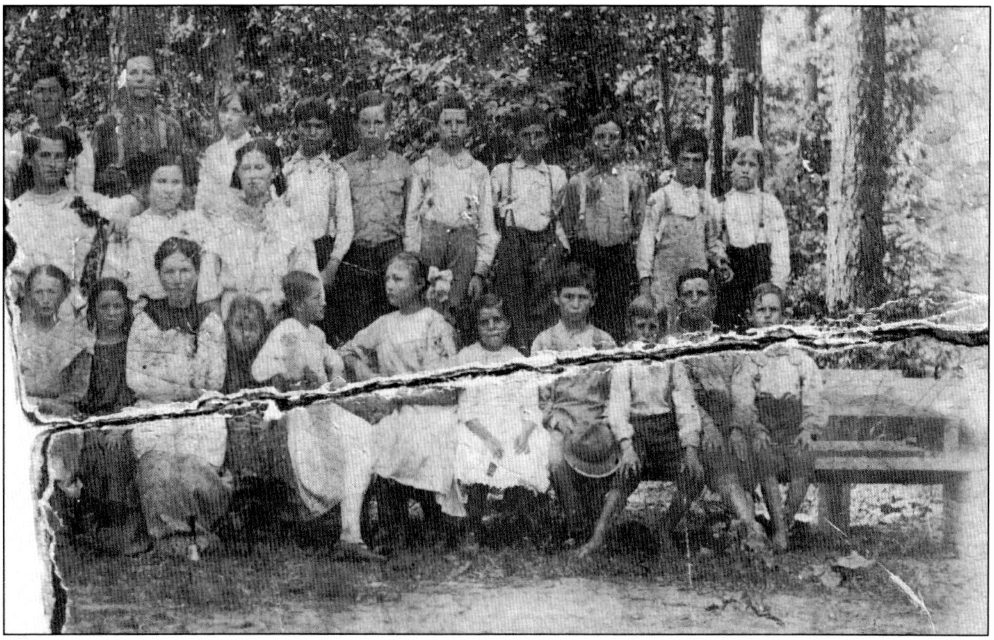

This 1910 photograph shows a Bois d'Arc School group. Shown here are, from left to right, the following: (first row) Lou Edward Wilson, Dovey Wilson, Mrs. Eddie Phillips (teacher), Susie Linley, Blanche Ophelia "Midge" McGarity, Florence Leatherwood, Ollie Stubbs, ? Harris, Luther Koon, Walter Koon, and Ivie Harris; (second row) Belle Rhodes, unidentified, and Dan Rhodes (twin of Belle); (third row) Clay Moring, Lewis Cox, two unidentified, James Harris, unidentified, Van Stubbs, James Cox, Acie Moring, and Rufus Berry. (Courtesy of the author.)

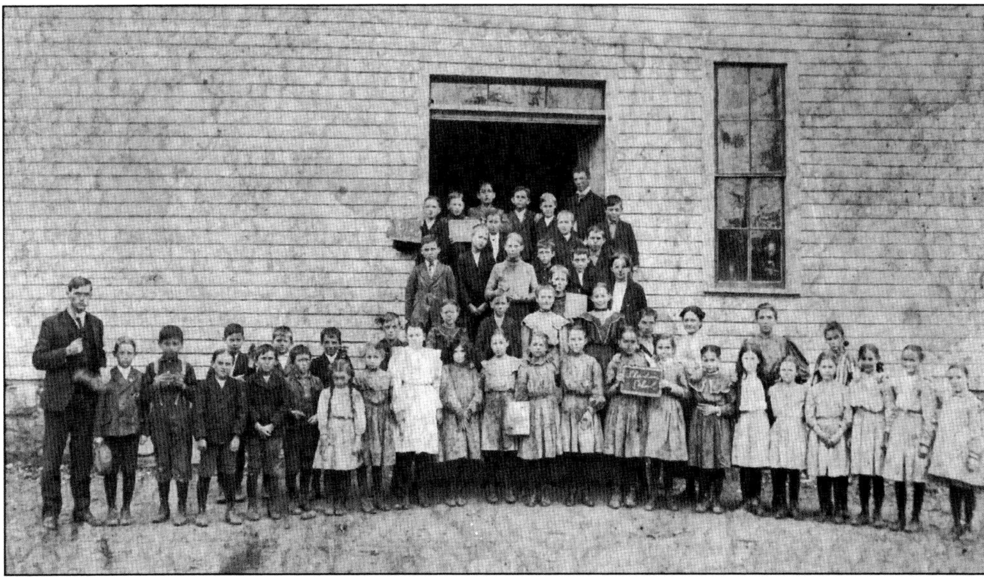

Sheridan School students pose for a photograph in the first decade of the 20th century. The school was located on the hill north of the present-day Sheridan Middle School campus. "Mattie Shepherd" is written on the reverse. (Courtesy of Grant County Museum.)

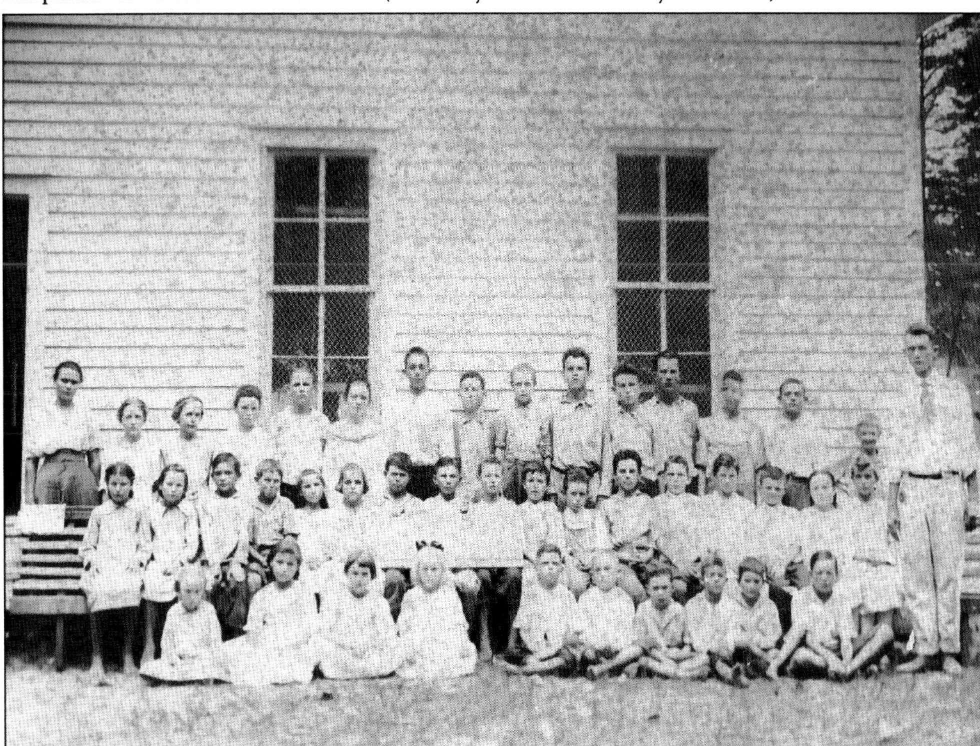

Included in this Little Creek School group, from the first decade of the 20th century, are Elmo Posey (second row, third from right) and Dalton Walker (second row, fifth from right). A school in this community was the meeting place for commissioners Allison, Harrison, and Morris, who were given the task of creating Grant County in 1869. (Courtesy of Grant County Museum.)

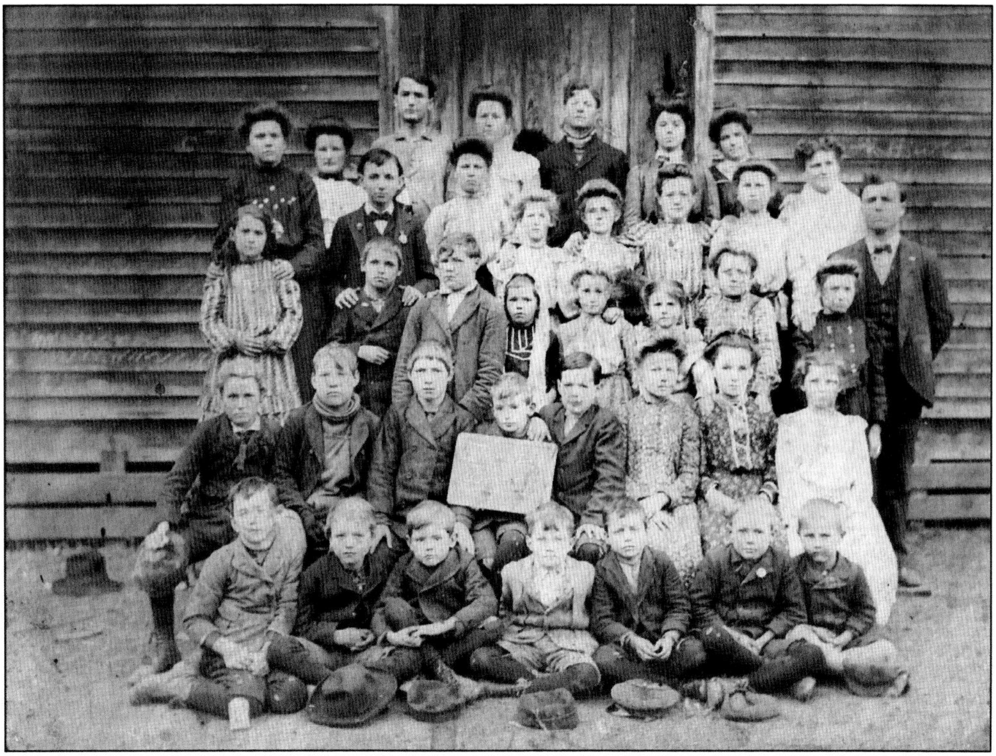

Shiloh School students assemble with a sign reading "December 16, 1904 District 17." Seated at left in the second row is Lawrence Wilson (born February 28, 1893), and May Wilson (Lawrence's first cousin) is sixth from left in the second row. May's father, Lafayette, left her as a small child, and she was raised by Jeff and Ida Wilson. (Courtesy of Grant County Museum.)

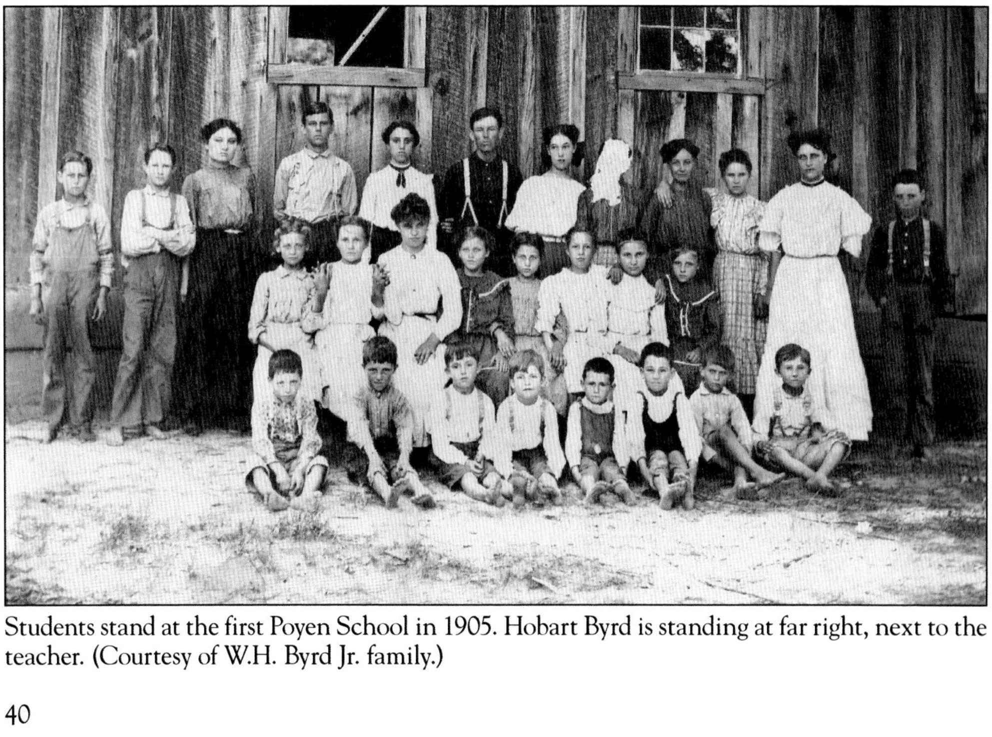

Students stand at the first Poyen School in 1905. Hobart Byrd is standing at far right, next to the teacher. (Courtesy of W.H. Byrd Jr. family.)

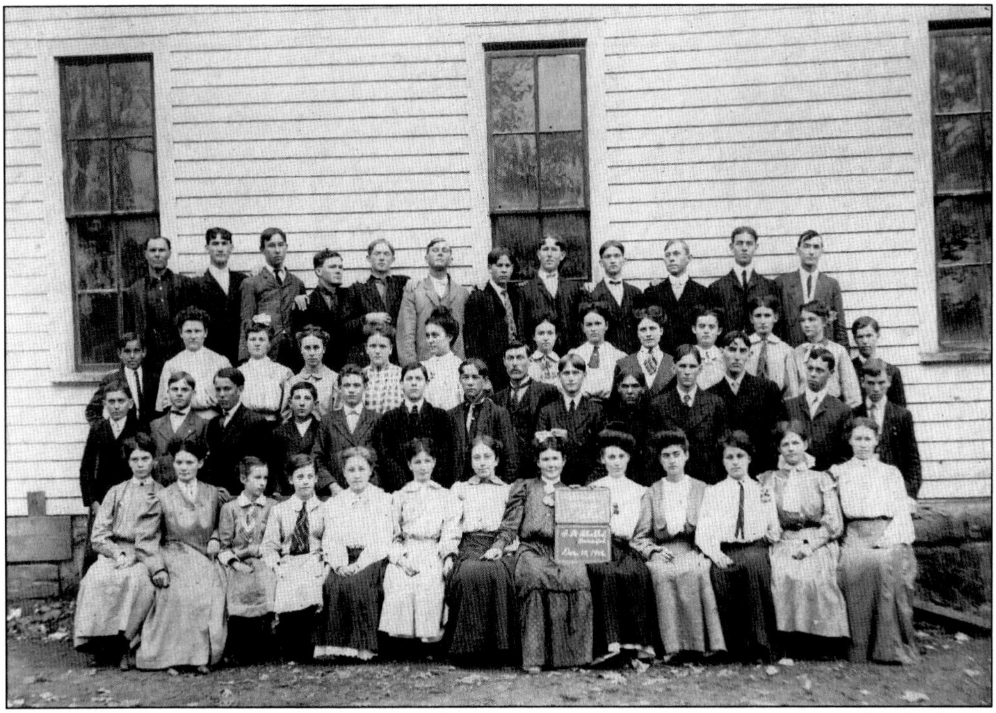

Sheridan School students posed for this photograph on December 10, 1906. I.W. Blacklock was principal of the school, located on the north side of the present-day Sheridan Middle School campus. Mattie Shepherd is in the third row, third from the right. In the fourth row, Sam McDonald stands fourth from left, and Oris Rushing is fourth from the right. On the back of this photograph are written the names "Mattie Shepherd" and "Earlene Shoptaw." (Courtesy of Grant County Museum.)

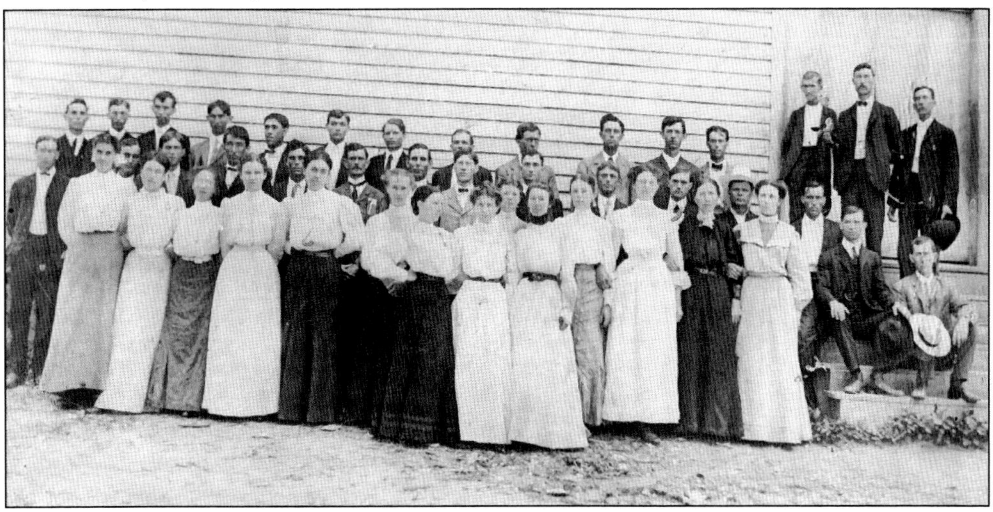

Civil War veteran and teacher trainer R.M. Rodgers stands at left in the doorway of the Sheridan School building, which served until the brick 1913 building was constructed about a block south. Oris Rushing is standing in the second row, third from left. "Local Links" in the December 13, 1900, *Sheridan Headlight* reported: "Prof. R.M. Rodgers informs us that 126 pupils have enrolled in the public school at this place." (Courtesy of Grant County Museum.)

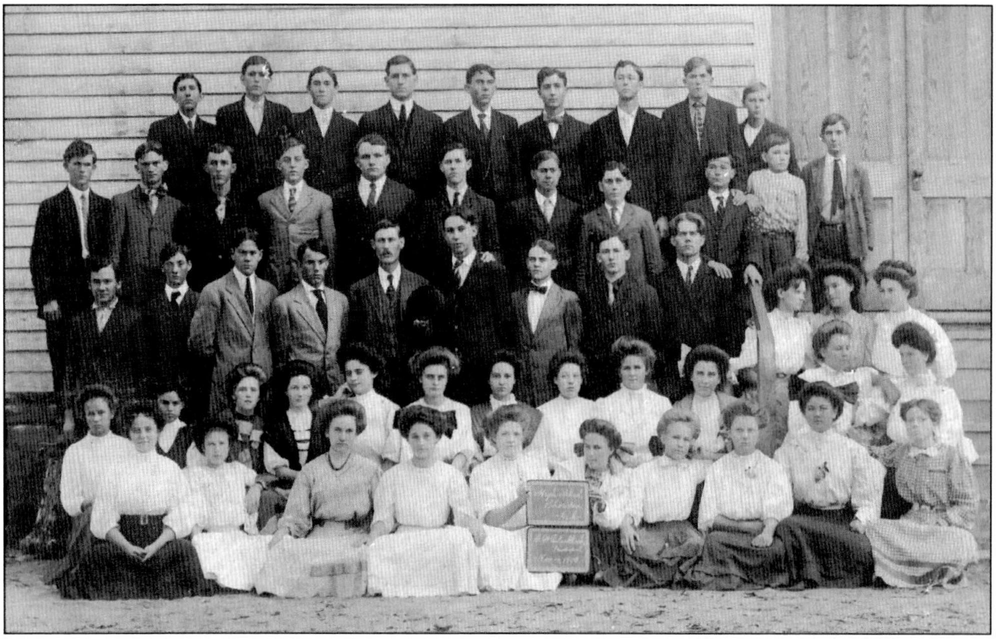

Sam McDonald was the first graduate honored with his name on the sidewalk in front of the 1913 brick school. In this photograph, taken in front of the original school building on School Street, he stands at far left in the third row. The slate reads "November 4, 1908, I.W. Blacklock, Principal, High School, Sheridan, Ark." This photograph was taken in an era when ladies used fake hairpieces, called "rats," which lifted their hair into the latest style. (Courtesy of Grant County Museum.)

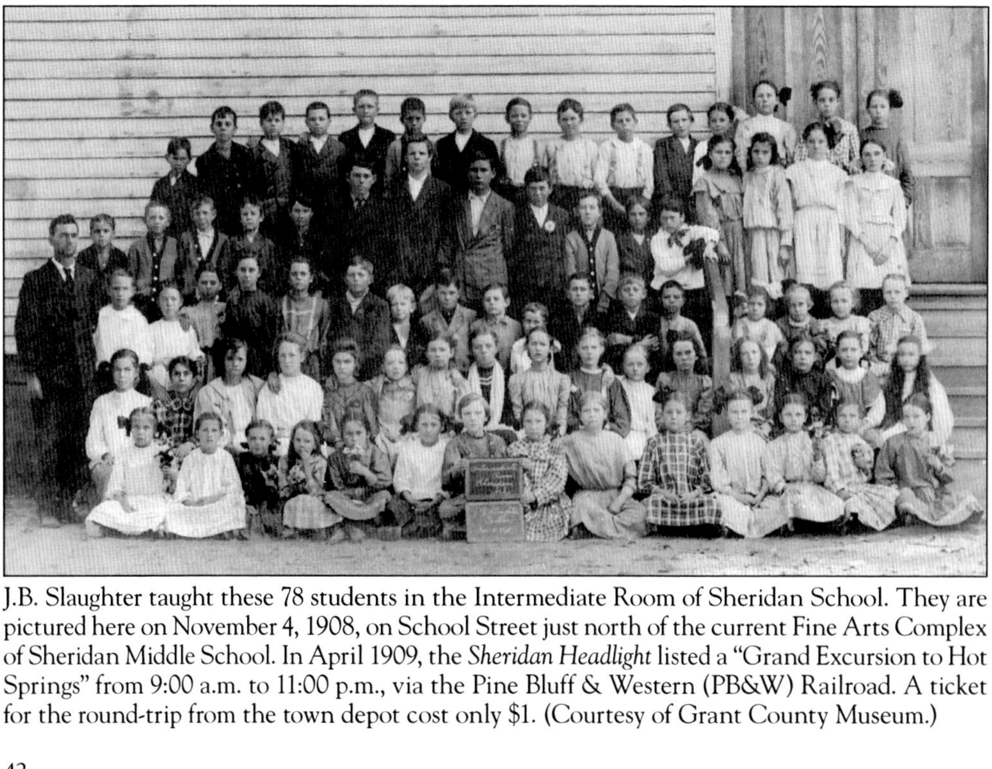

J.B. Slaughter taught these 78 students in the Intermediate Room of Sheridan School. They are pictured here on November 4, 1908, on School Street just north of the current Fine Arts Complex of Sheridan Middle School. In April 1909, the *Sheridan Headlight* listed a "Grand Excursion to Hot Springs" from 9:00 a.m. to 11:00 p.m., via the Pine Bluff & Western (PB&W) Railroad. A ticket for the round-trip from the town depot cost only $1. (Courtesy of Grant County Museum.)

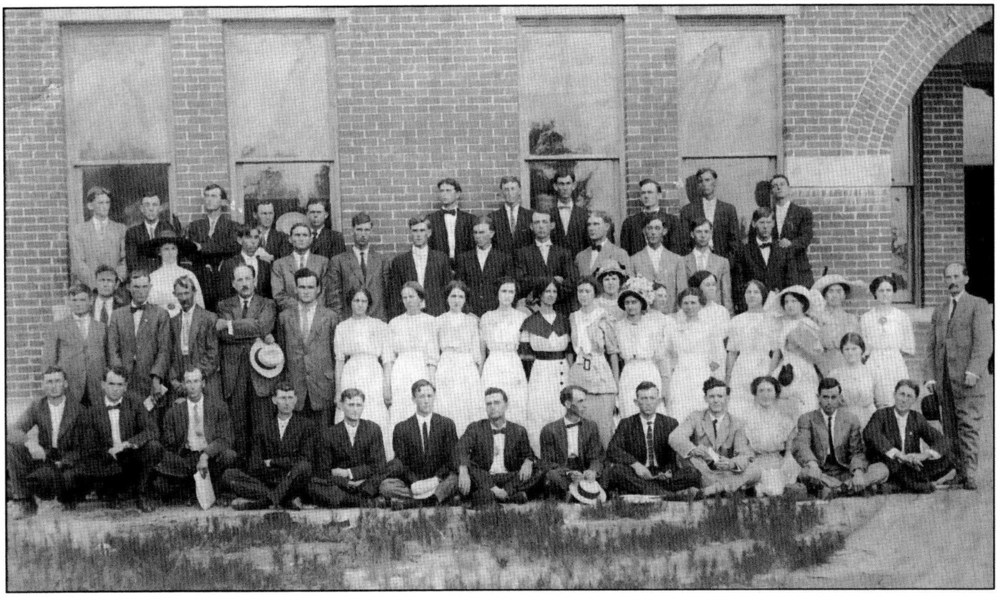

Grant County's teacher association met at the courthouse. Included in this photograph are the following: (first row, from left) Eric Ross (5th), Nathan Nall (6th), Benton Slaughter (7th), Arthur Cearley (9th), and Ed F. McDonald (10th); (second row, from left) Sam McDonald (5th), Etta Slaughter (10th), Lala Mathews (12th, in hat with stars), Alma Rodgers (13th, partially hidden), and Mana Harrison (3nd from right, with large hat); (third row from left) ? Breazeal (4th), Bob Mathews (6th), and Charlie Mathews (9th, in bow tie); (fourth row) Oris Rushing (far left) and R.L. Slaughter (far right). (Courtesy of Grant County Museum.)

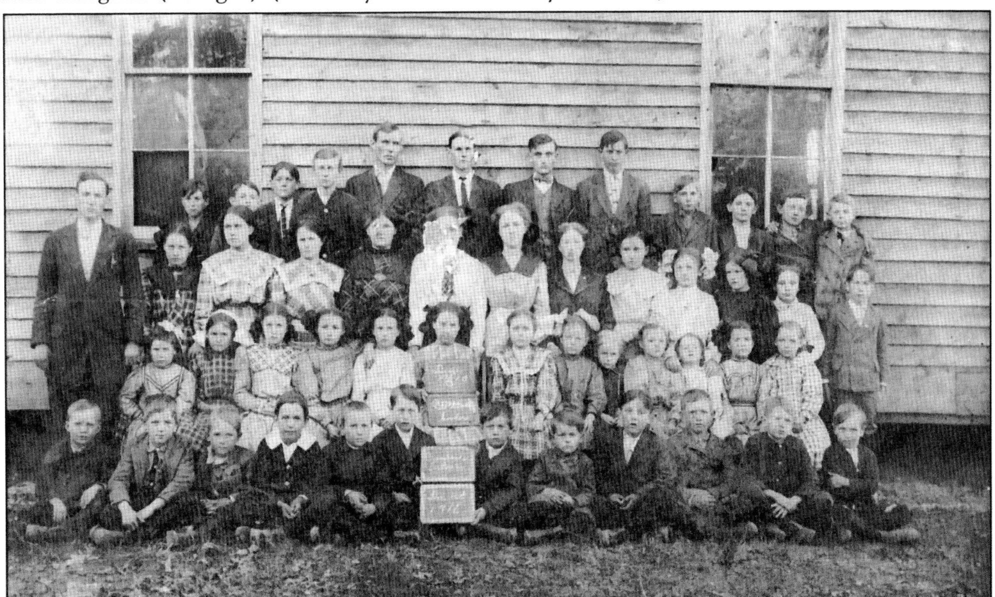

This photograph, taken on December 1, 1911, shows students of the Crossroads School, District Number 10, taught by C.F. McDonald. The building remains standing south of New Hope Methodist Church. Upstairs was the meeting hall of the Mooneyville Masonic Lodge. (Courtesy of Grant County Museum.)

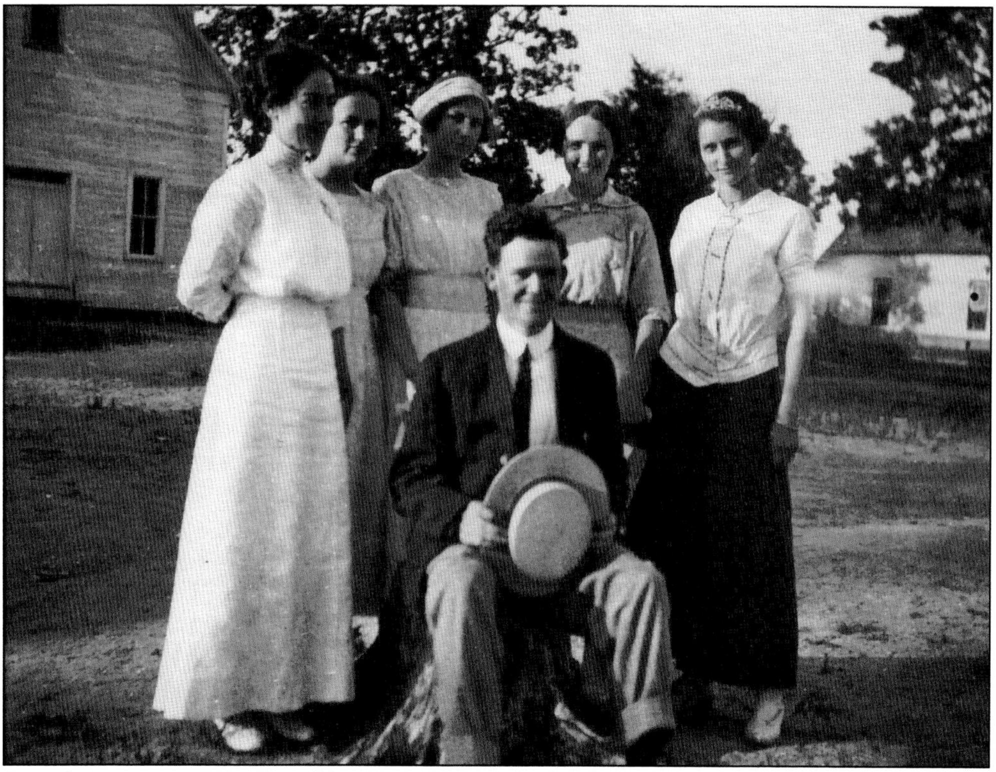

Seated on a stump, Ed McDonald poses with a group of early-20th-century Grant County teachers, including Mana Harrison (far left) and Delma Walker Rushing (third from left). The building behind them is an early Sheridan school, located on the appropriately named School Street. McDonald was principal of Sheridan High School in 1919 and 1920, later serving three terms as Arkansas's secretary of state. (Courtesy of Grant County Museum.)

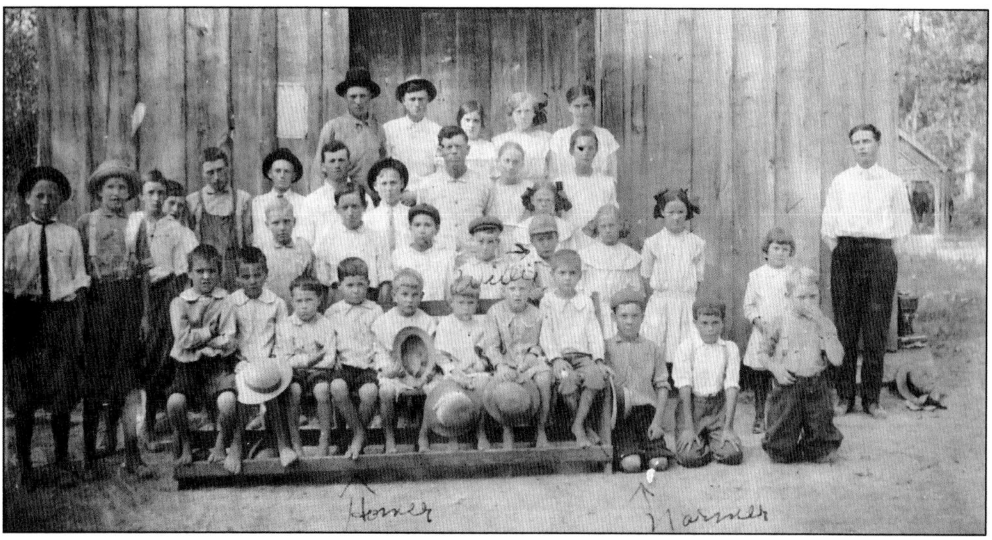

Among the Corinth School students in this 1912 photograph are Homer Wright (first row, fourth from left) and Normer Wright (first row, ninth from left). Teacher Walter Pettus is at the far right. (Courtesy of Grant County Museum.)

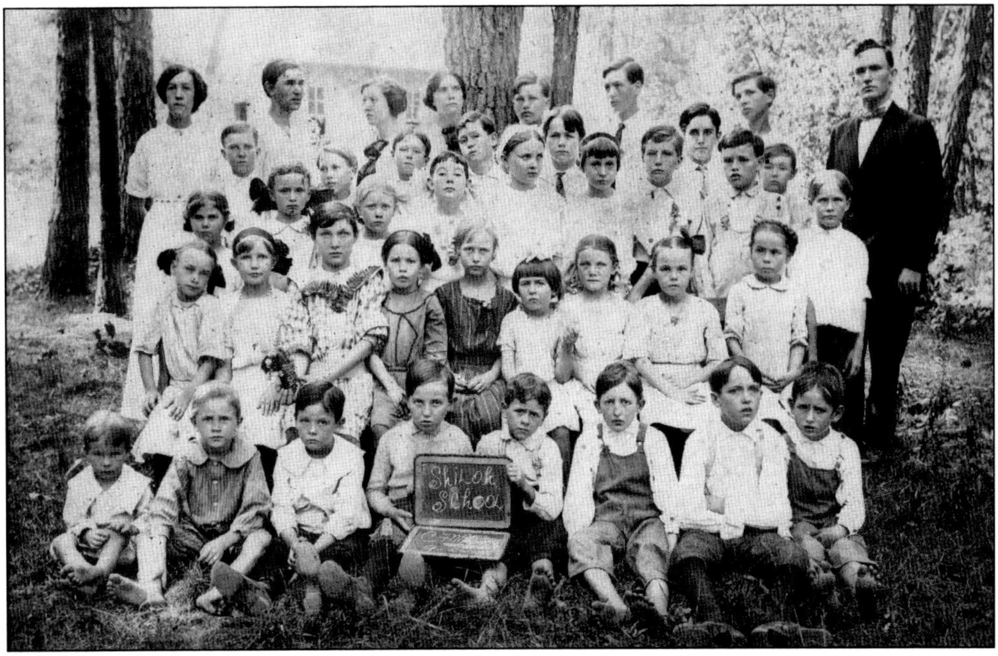


Shiloh School students pose for a photograph in 1913. Shown here are, from left to right, the following: (first row) Pud Crouse, Ivan Allen, Clyde Crouse, Olen Duncan, Alfred Barnes, Tom Barnes, Idus Barnes, and Calvin Barnes; (second row) Wilma Allen, Bertha Hyde, Tennie "Dutch" Ward, Mamie Reynolds, Neville Wilson, Inez Duncan, unidentified, Doris Crouse, and Mamie Womble; (third row) unidentified, Ray Barnes (large bow), Ola Henslee, Maudi Duncan, Ethel Henslee, Troy Shackleford, unidentified, Fay Barnes, and unidentified; (fourth row) Houston Crouse, Ralph Wilson, Rayford Barnes, two unidentified, Otis Bratton, Martin Barnes, and teacher Cliff McDonald; (fifth row) Lela Bratton, Fred Allen, Della Henslee, Alice "Big Sis" Henslee, Davis Wilson, Earl Barnes, and Lon Hyde. (Courtesy of the author.)

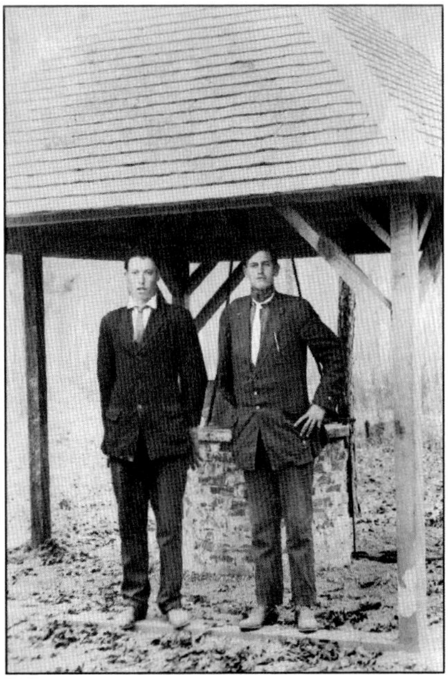

Hobart Byrd (left) and his cousin, George Pierce, stand near the well located east of the main Poyen School building. Byrd later participated in World War I as a medic. (Courtesy of W.H. Byrd Jr. family.)

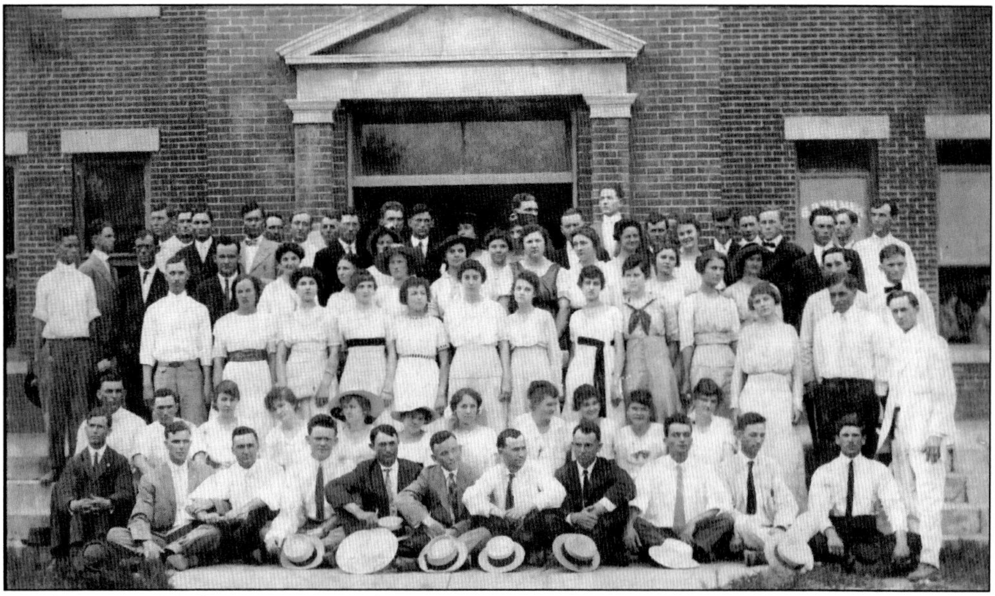

Teachers assemble for a meeting at the Grant County Courthouse in 1914. Oris Rushing, in the third row at far right wearing a bow tie, later became county school superintendent and was responsible for starting the progressive idea of consolidating the local one-room schools in the late 1920s. He would personally take teachers to Monticello for additional college coursework. (Courtesy of Grant County Museum.)

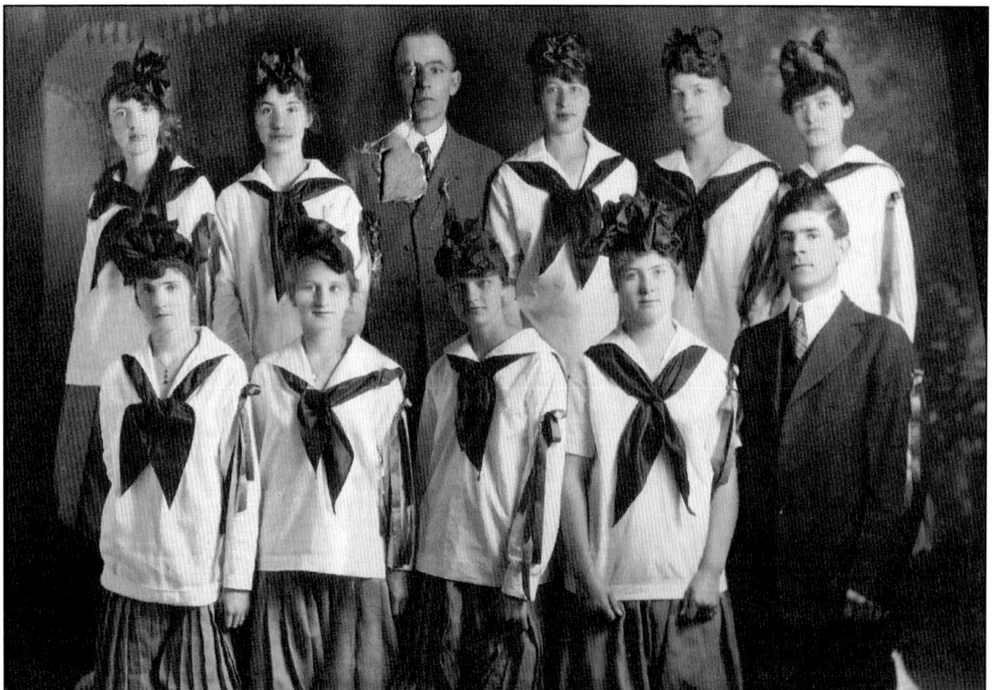

The Sheridan School girls' basketball team poses for a photograph in 1914. The coach, Marvin Harris, is at lower right. Sports uniforms were obviously much different. (Courtesy of Grant County Museum.)

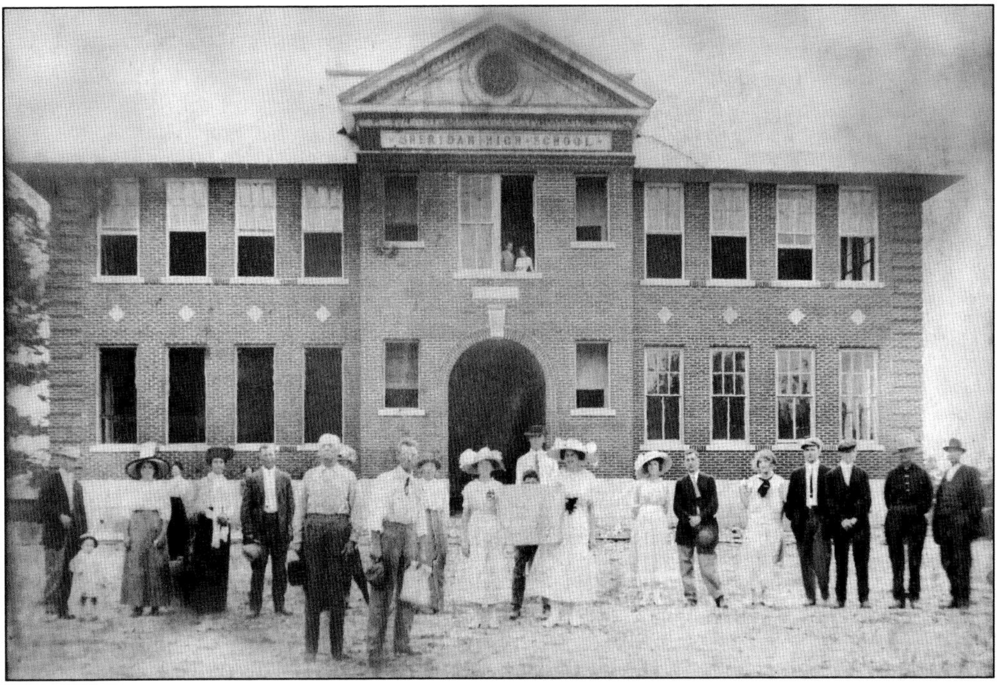


The first brick school in Grant County was constructed on the former site of the McDonald family's cornfield. In this photograph, city residents stand in front of the almost-completed project. The man holding blueprints may be Clyde A. Ferrell, the architect. School board members included J.L. Butler, president (standing to Ferrell's right); T.C. Rushing, secretary; I.S. McClellan; F.K. Marks; C.D. Harris; and G.W. Baker. Shown standing at right are Isaac McClellan (second from right), Oris Rushing (third from right), and John L. McClellan (fourth from right), who became a US senator. (Courtesy of Lois Lea Clark and Grant County Museum.)

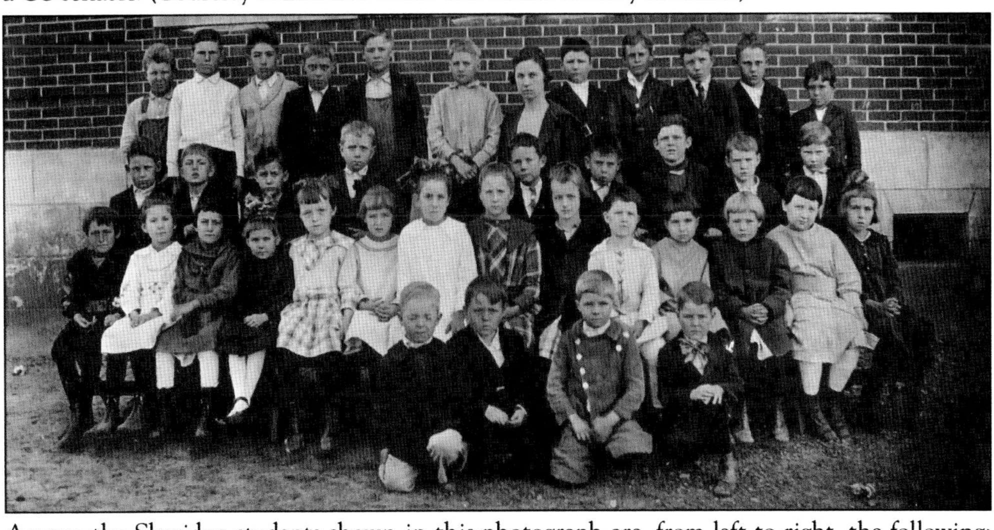

Among the Sheridan students shown in this photograph are, from left to right, the following: (first row) Edward Lee McDonald, Utley Waddell, Jack Baker, and Eldred Gean; (second row) Rebecca Monroe, Mary Wilson, Essie Koon, Glennie Mae Mingea, Retha Jones, five unidentified, Juanita McCool, Eunice Swafford, unidentified, and Lillean Hill. No names were listed for the back two rows. (Courtesy of Doug Lloyd.)

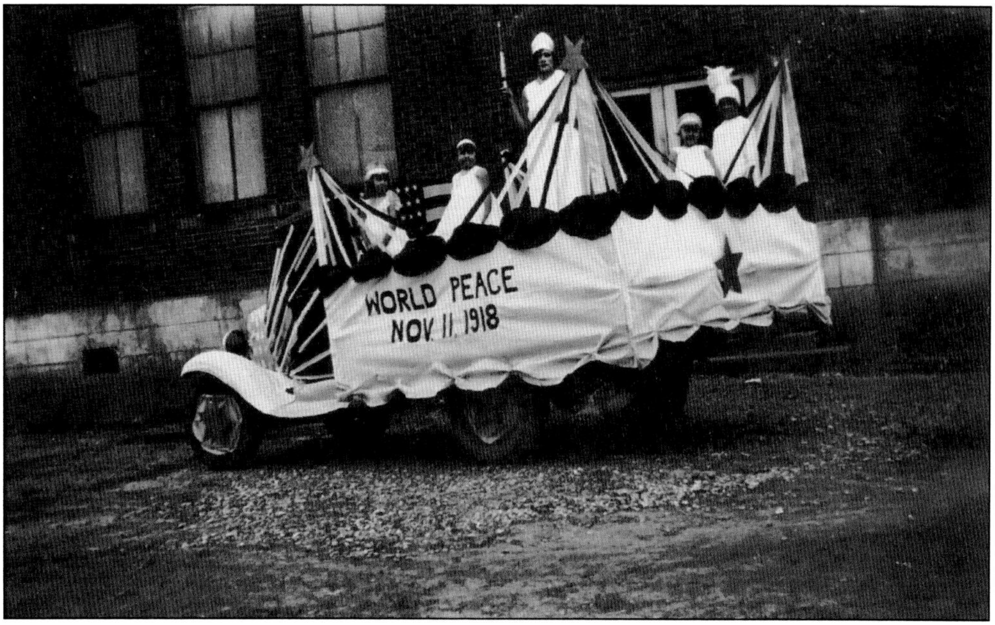


Parked beside the 1913 Sheridan School building is the float that won first prize, celebrating the end of World War I on November 11, 1918. (Courtesy of Doug Lloyd.)

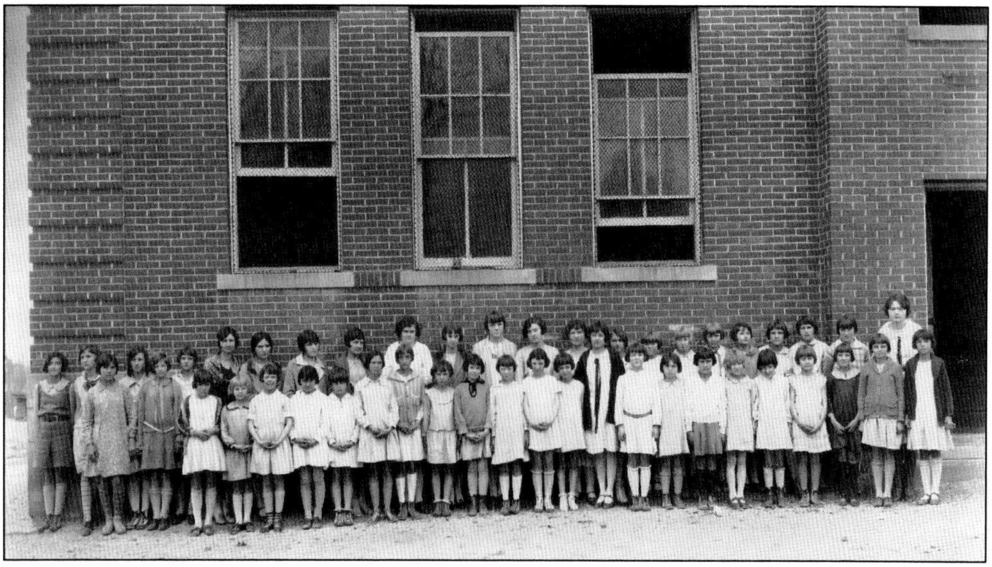

Sheridan School's music department poses for this 1920s photograph. Shown here are, from left to right, (first row) Norma Stevenson, Ethel Nelson, Wanda Dodd, Sibyl Williams, Sarah Henslee, Mildred Bradley, Margie Nall, Sibyl Lewis, Lorene Stuckey, Doris McClellan, Katherine Nelson, Estelle Howard, Anna Sims, Louise Hope, Essie Koon, Emerine Toler, Ezelle Freeman, Clara Griffin, Reba Hood, Reba Childers, Gladys Reaves, Cora Bratton, Floy Vanlandingham, and Bernice Ashcraft; (second row) Gladys Ashcraft, Mary Kelly, Minnie Murry, Venita Stuckey, Hazel Mathews, Helen Jordan, Odessa Stuckey, Ruby Everette, Reba Carr, Bernice Mathews, Retha Jones, Ruby Koon, Lois Ferguson, Charlene Mathews, Doris Shepherd, Juanita Holiman, Betty Glassco, Henriette Everette, Lorene Goolsby, Irma Sumners, Imogene Holiman, and Lexia Jones. (Courtesy of Doug Lloyd.)

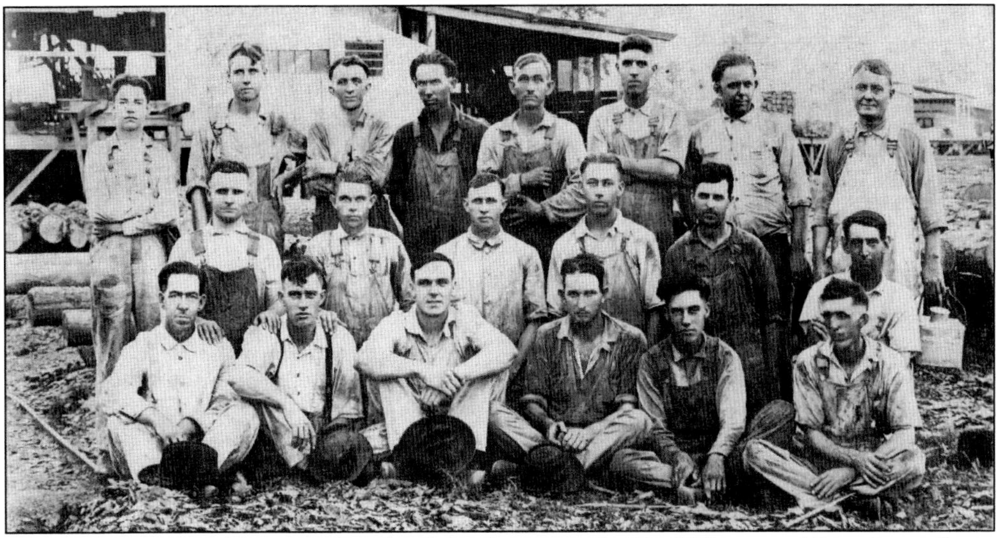

Located to the east of the 1913 Sheridan School campus, the J.L. Williams and Sons Sawmill was built in the 1920s. The mill was the chief employer of county citizens for over four decades. The Williams's Oak Flooring Mill became a national leader, shipping a large order for the remodeling of the White House in Washington, DC, in the early 1950s, according to Oliver Williams, son of the mill's owner. (Courtesy of Grant County Museum.)

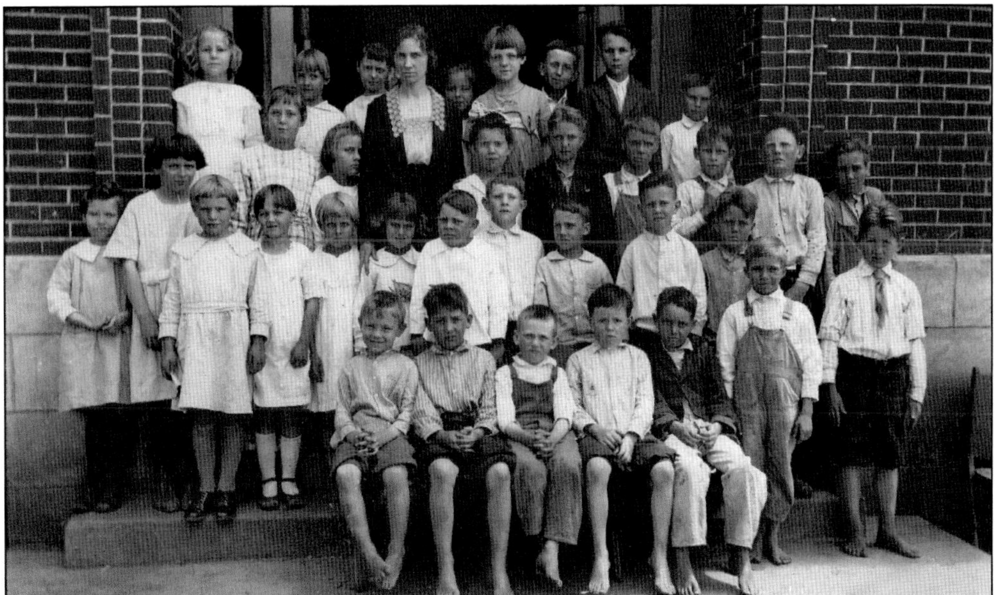

A Sheridan School class poses for a photograph. Shown are, from left to right, (first row) Cloyd Bratton, Welles Avery, unidentified, Loren Worthen, Utley Waddell, Jack Baker, and Willie Pate; (second row) unidentified, Vita Reid, Eunice Swafford, unidentified, Leola Knowles, Bernice Mathews, Carol Henslee, and four unidentified; (third row) Selma Greenlea (in dress with squares), Geneva Love, teacher Delma Walker Rushing, Reba Stevenson, Bob McCool, Pod Baker, and three unidentified; (fourth row) Reba McCool, Mildred Swiney, Mildred Harper, unidentified, Retha Jones, and three unidentified. The photographer was from the Murtik Company of Pine Bluff. (Courtesy of Doug Lloyd.)

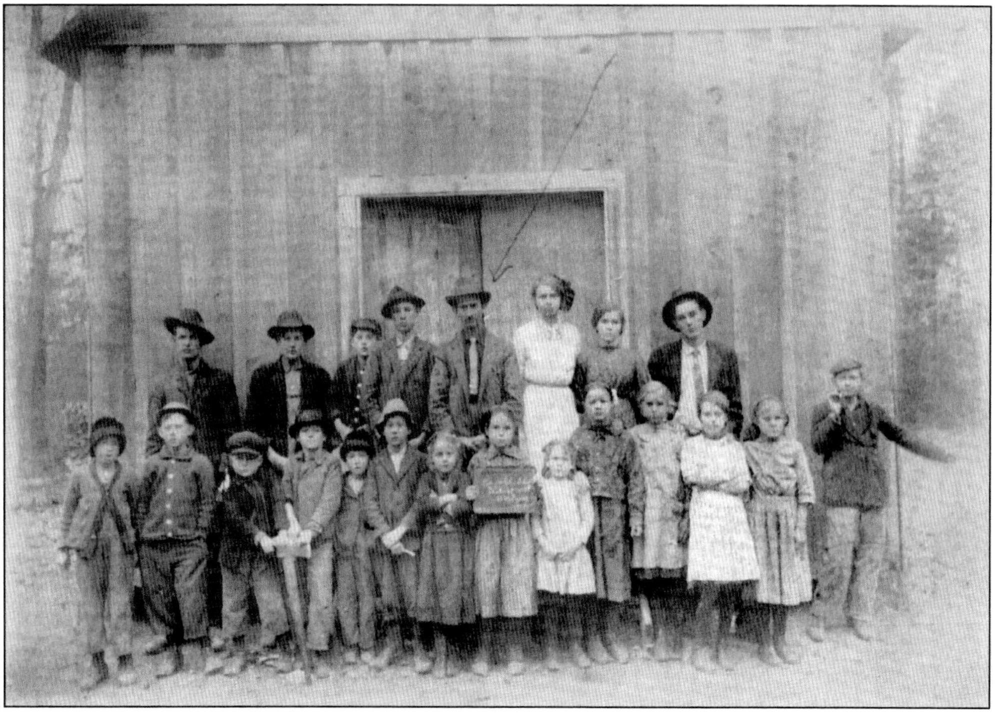

Located south of Tull in northwestern Grant County, the Frog Level School served area pupils in search of an education. The teacher is Will DuVall (second row, center). (Courtesy of Grant County Museum.)

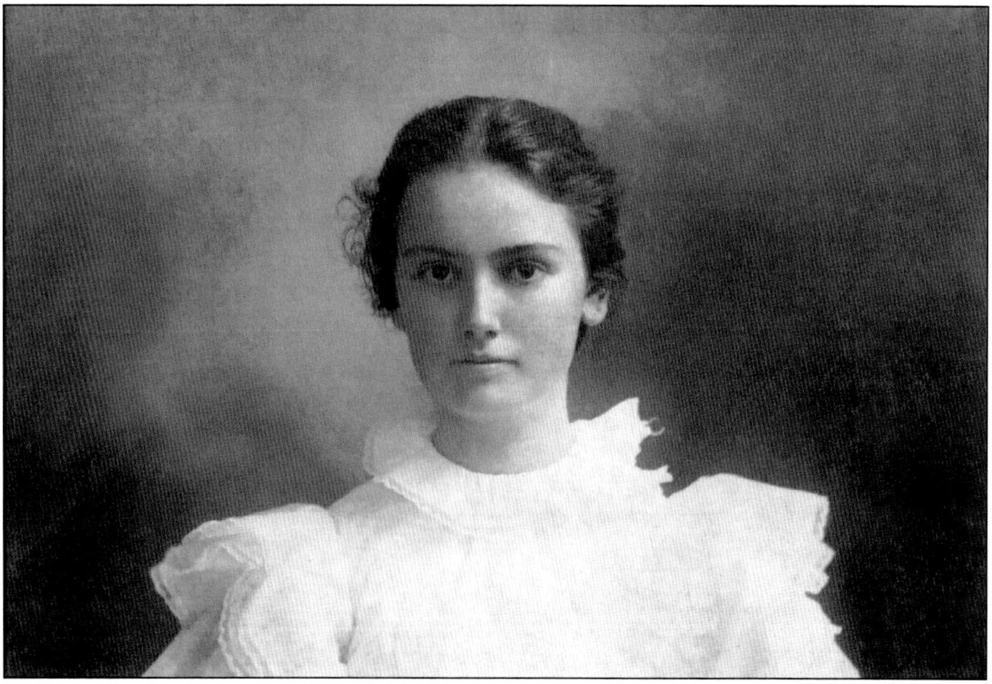

Striking a lovely pose in this photograph is Nora Harrison of Leola. She and her siblings enjoyed lives of achievement. (Courtesy of Grant County Museum.)

This 1920 invitation, bound with a green cord, summoned friends to the graduation of 14 Sheridan High seniors. (Courtesy of Doug Lloyd.)

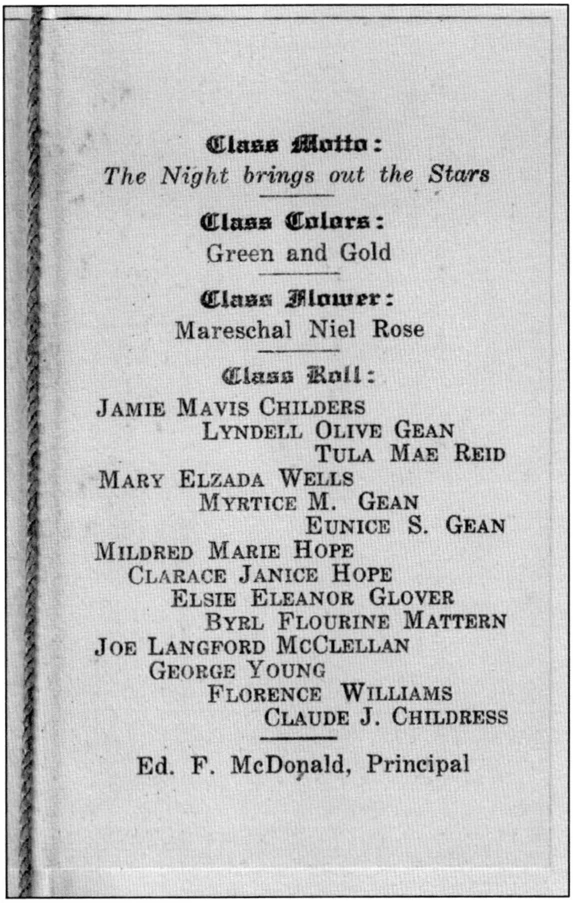

**Class Motto:**
*The Night brings out the Stars*

**Class Colors:**
Green and Gold

**Class Flower:**
Mareschal Niel Rose

**Class Roll:**
JAMIE MAVIS CHILDERS
LYNDELL OLIVE GEAN
TULA MAE REID
MARY ELZADA WELLS
MYRTICE M. GEAN
EUNICE S. GEAN
MILDRED MARIE HOPE
CLARACE JANICE HOPE
ELSIE ELEANOR GLOVER
BYRL FLOURINE MATTERN
JOE LANGFORD MCCLELLAN
GEORGE YOUNG
FLORENCE WILLIAMS
CLAUDE J. CHILDRESS

Ed. F. McDonald, Principal

Embossed with the school's Yellowjacket emblem, this card invited guests to the 1929 Sheridan School graduation. (Courtesy of Doug Lloyd.)

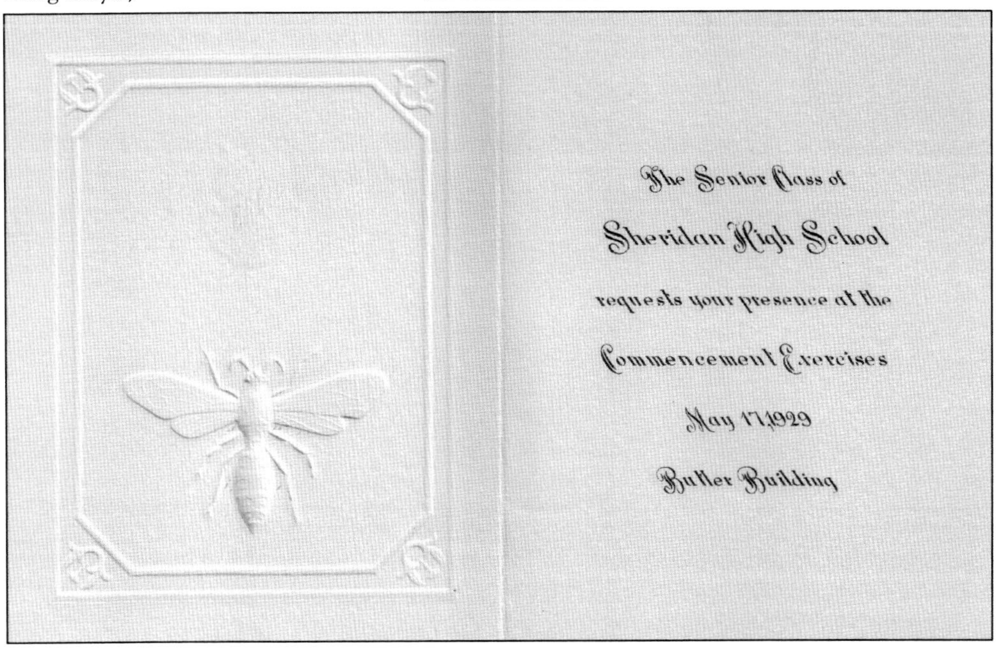

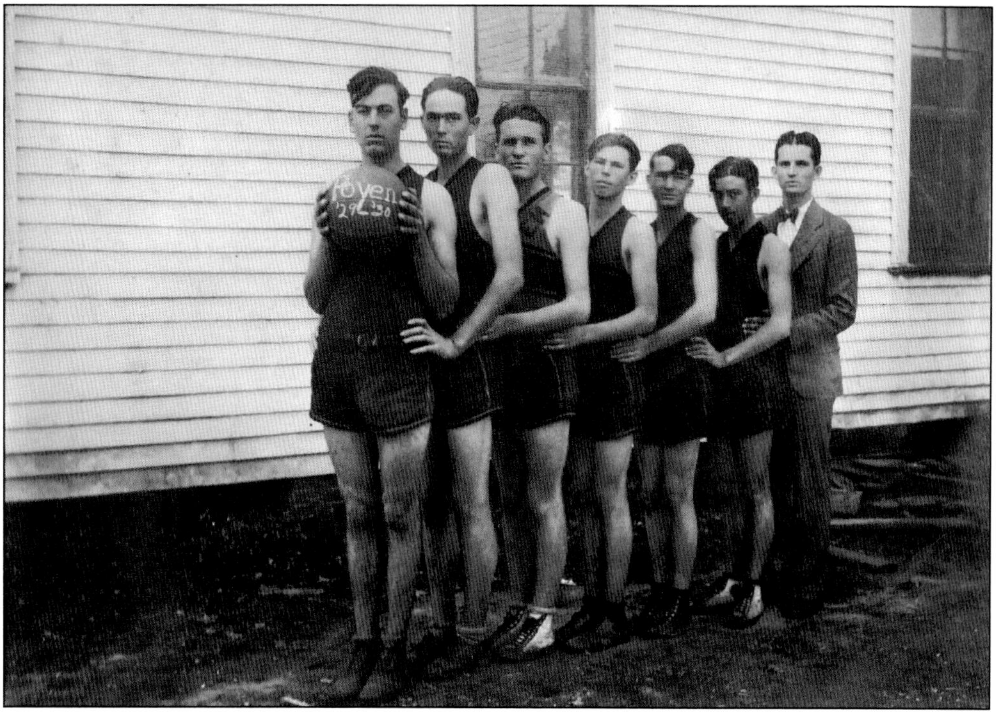

Arvin Byrd, at left, proudly holds the Poyen School team basketball. During World War II, Byrd delivered supplies by mule train in Burma and China. (Courtesy of W.H. Byrd Jr. family.)

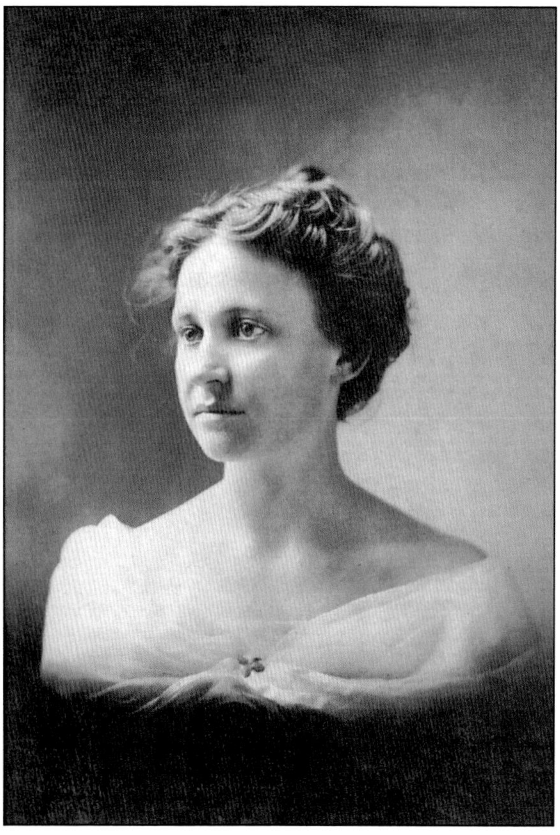

Beauty and grace are evident in this photograph, found in the Dr. John W. Harrison collection. (Courtesy of Grant County Museum.)

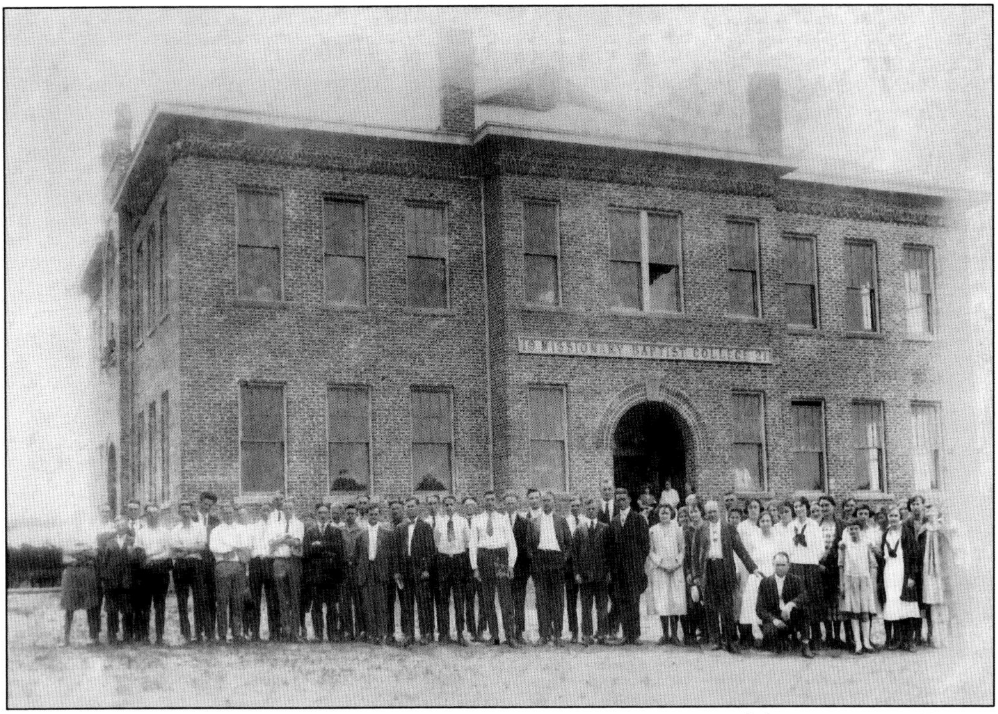

In this photograph, citizens of Sheridan celebrate the construction of the Missionary Baptist College on Church Street. Built in 1921 and similar in design to the 1913 Sheridan School, the college offered additional coursework for students. It closed in 1934, and the property became an elementary school. (Courtesy of Grant County Museum.)

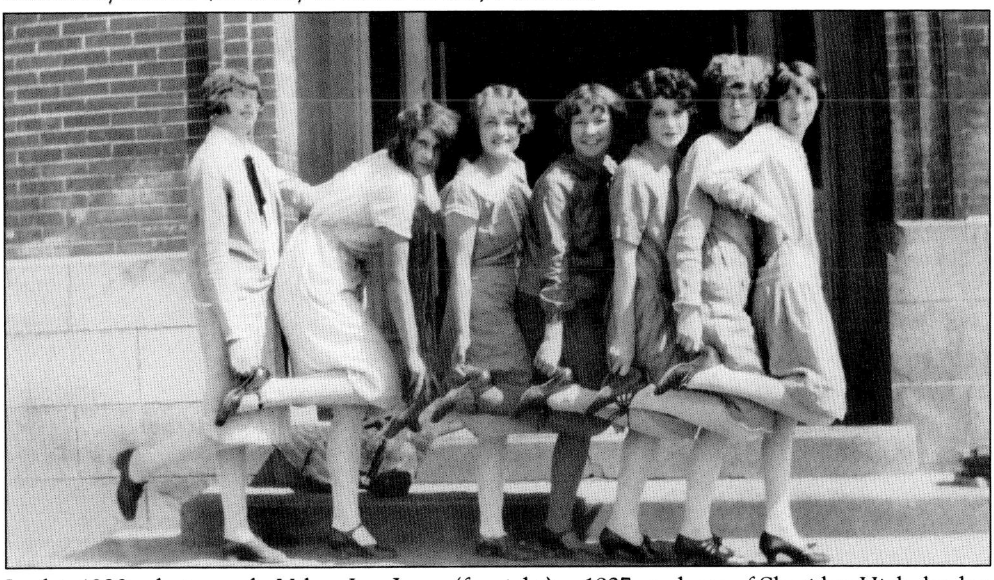

In this 1920s photograph, Velma Lea Jones (far right), a 1927 graduate of Sheridan High, leads a group of girls posing at the 1913 school building. In the summer, her parents, Lem and Claudia Reese Jones, permitted Velma and her sister Retha to experience the Joyzelle camp for girls near Monte Ne, now under Beaver Lake. Velma later married fellow graduate Joseph Dee Lybrand. (Courtesy of Doug Lloyd.)

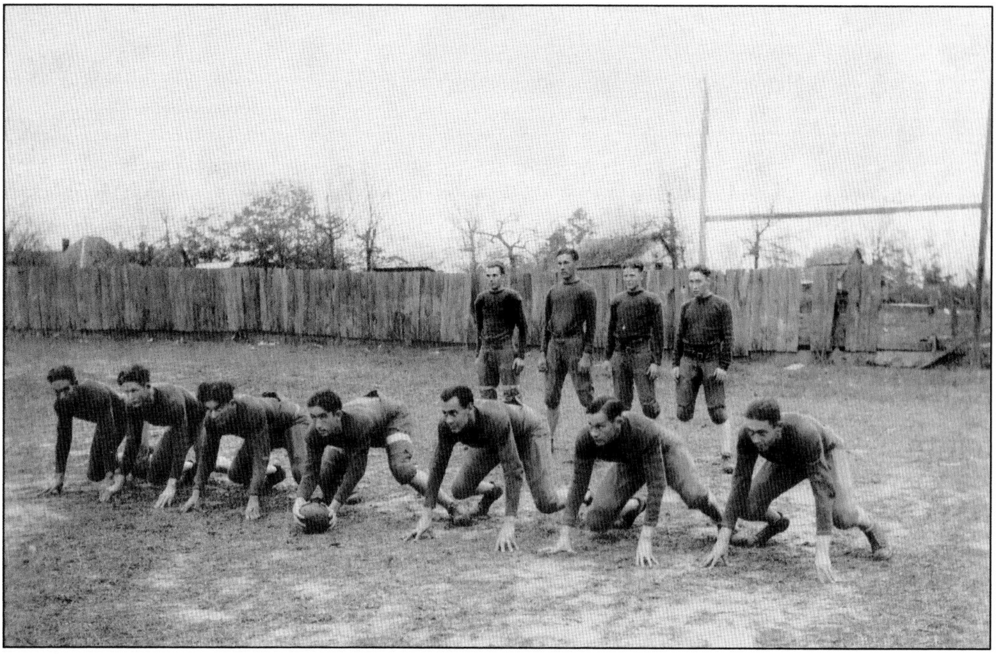

On the original Sheridan football field, members of the 1926 team pose for a photograph. Shown here are, from left to right: (first row) Hollis Mathis, Lee Drake, Iro Keesee, Glen Ashcraft, Otho Morris, Clyde Koon, and unidentified; (second row) Herman Knowles, Frank Koon, Macon Gean, and Red Varnell. (Courtesy of Doug Lloyd.)

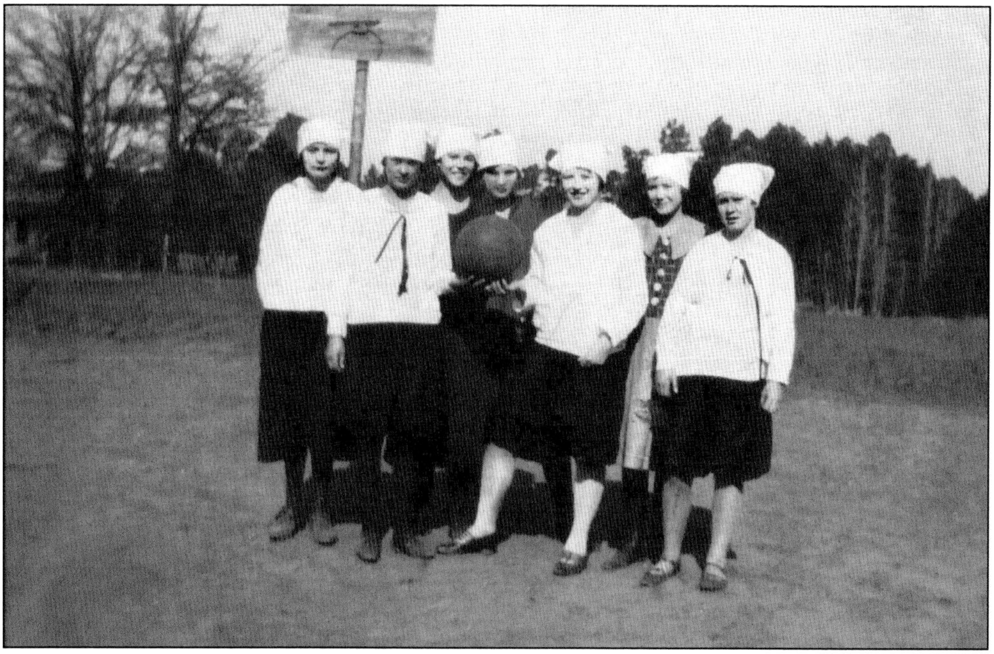

Dressed in homemade team uniforms, the Shiloh School girls' basketball team is pictured in the early 1920s. The team members are, from left to right, Inez Duncan (Caldwell), Wilma Allen (Ross), Merle Henslee (McCool), May Adams (English), Mexer Adams (Crouse), Jessie Lee Henslee (Young), and Cloma Ashcraft (Wiggins). (Courtesy of the author.)

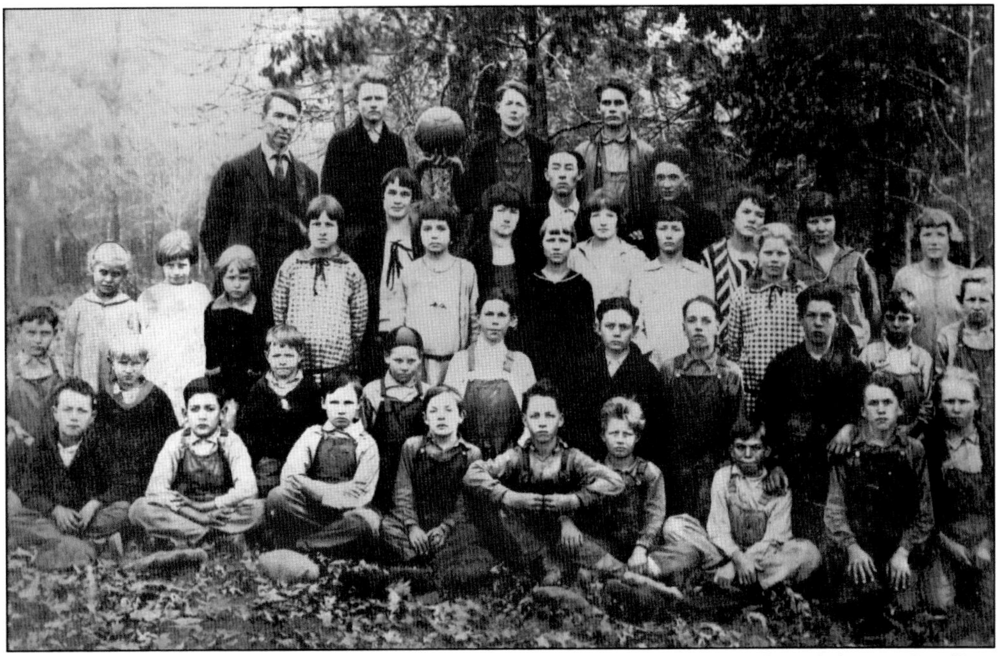

A Shiloh School class poses for a photograph in 1926. Shown here are, from left to right, (first row) Tootie Allen, Olen Acree, Raymond Crouse, Boyce Allen, Buck Allen, Gus Adams, Herman Womble, Jewel Allen, and Walter Adams; (second row) Cleo Camp, Garlin Stout, Leon Allen, Ponny Crouse, Pud Crouse, Gus Duncan, Doyle Allen, Jess Hunter, DeVoy Womble, and William Barnes; (third row) Nina Womble, Jessie Adams, Hazel Allen, Gracie Lou Stout, Wilma Allen, Ruby Allen, Myrl Allen, Alma Wilson, Mexer Adams, Estelle Barnes, Merle Henslee, Grace Adams, Mac Adams, and Jessie Lee Henslee; (fourth row) teacher Ed McCool, Carl Wilson, Edward Barnes, Martin Barnes, Lewis Womble, and Ivan Allen. (Courtesy of Hazel Wilson.)

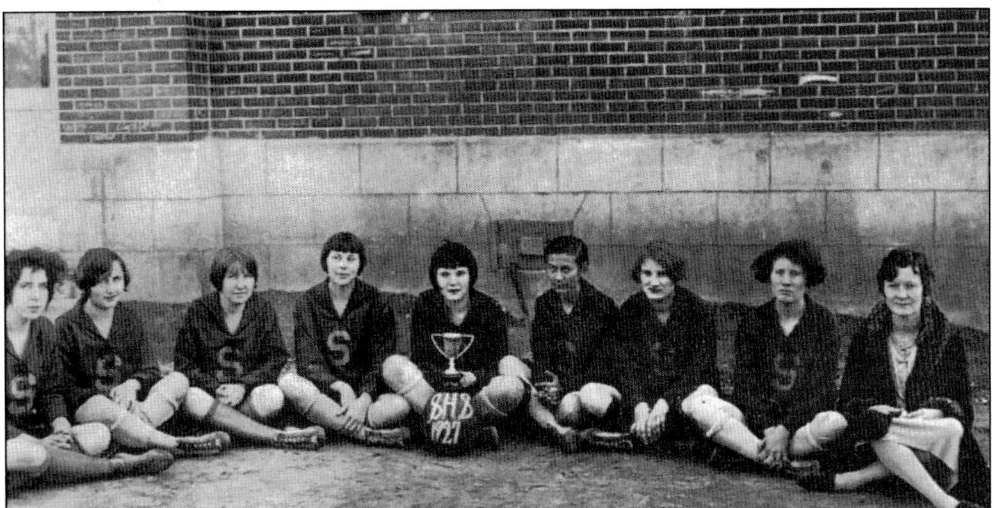


The 1927 Sheridan High School girls' basketball team proudly displays a trophy beside the 1913 school. Their team was organized in 1926, with Mrs. L.A. Francis as coach. School colors were royal purple and gold. In 1926, there were 360 students attending the Sheridan District, whose annual budget was $5,000. (Courtesy of Doug Lloyd and Grant County Museum.)

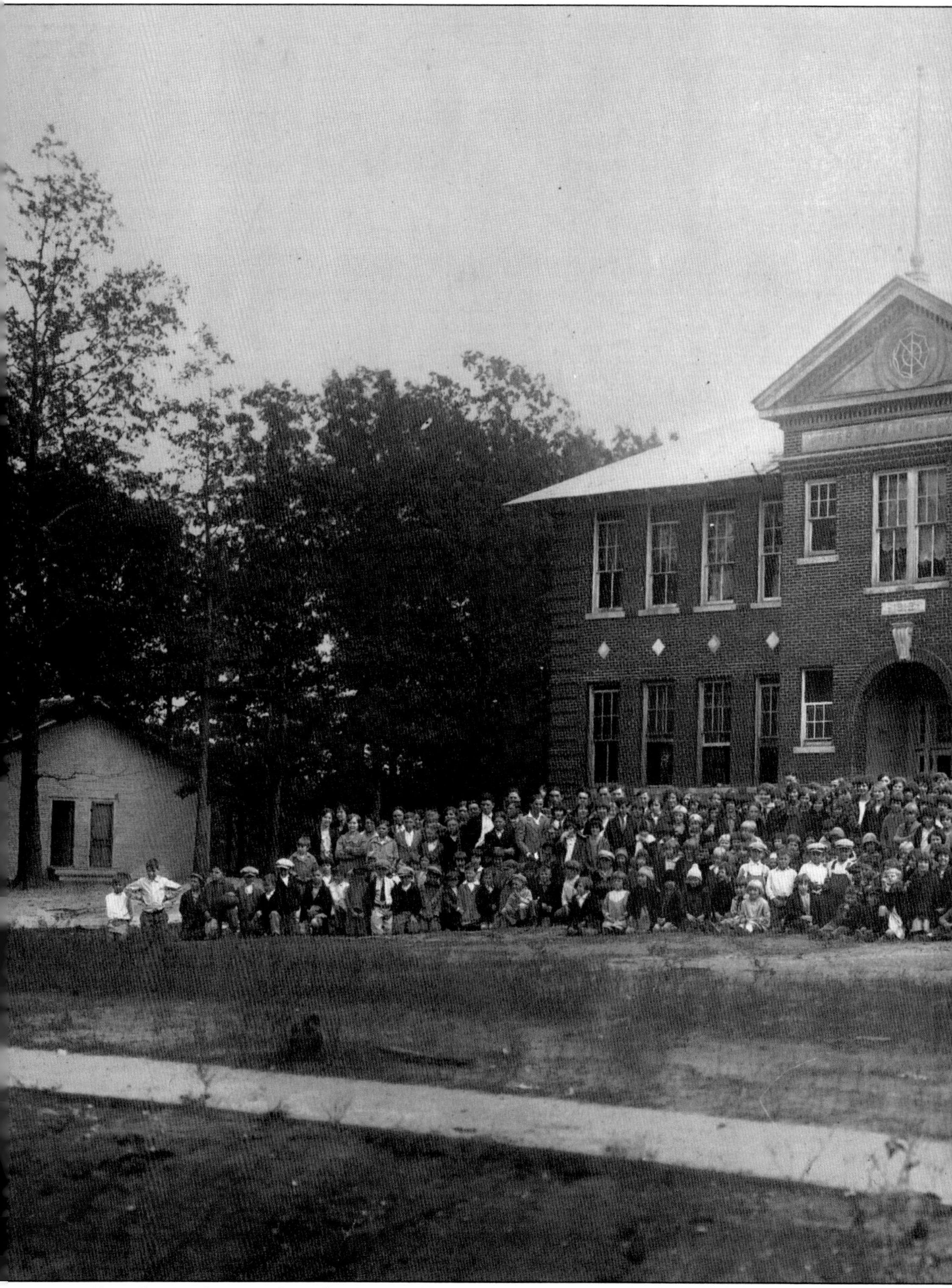

Sheridan students and teachers assemble in 1927 for a photograph in front of the 1913 school

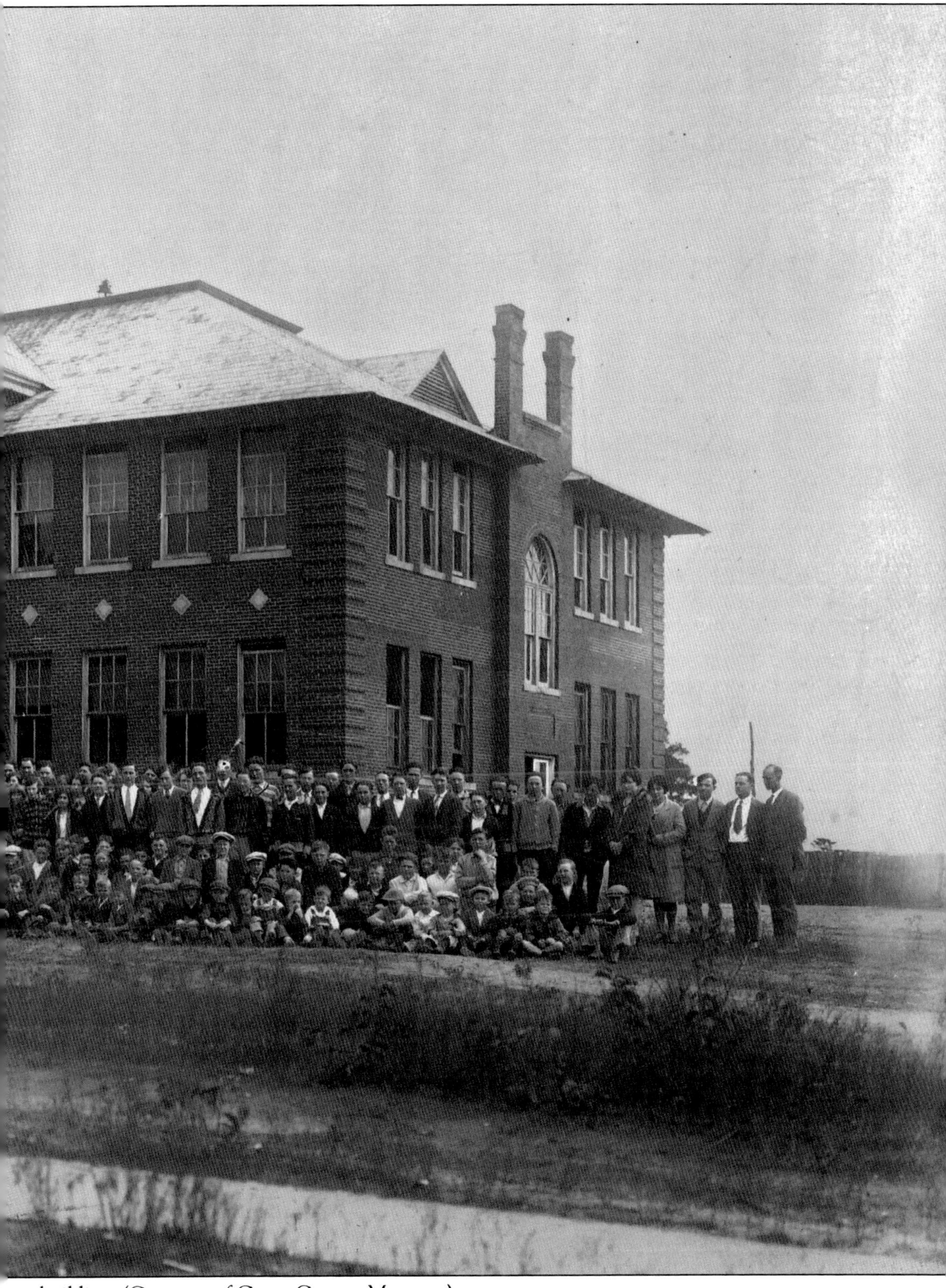

building. (Courtesy of Grant County Museum.)

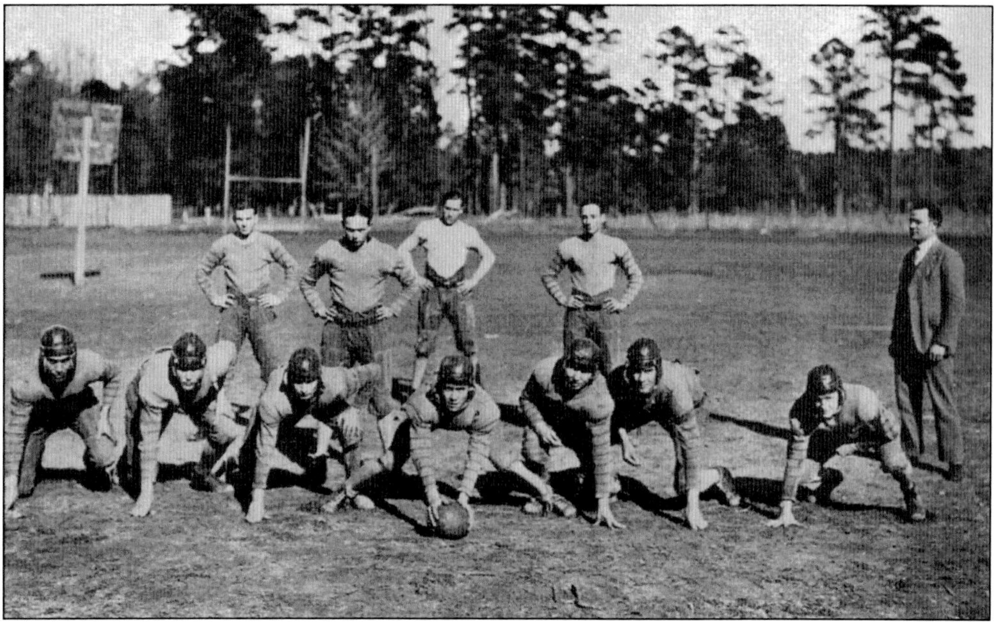

In this 1927 photograph, Sheridan Yellowjacket football players in leather helmets line up with coach and superintendent Lemuel A. Francis (far right) on the first football field, at what is currently the site of the Kelly Williams Gym of Sheridan Middle School. (Courtesy Sheridan School District yearbook.)

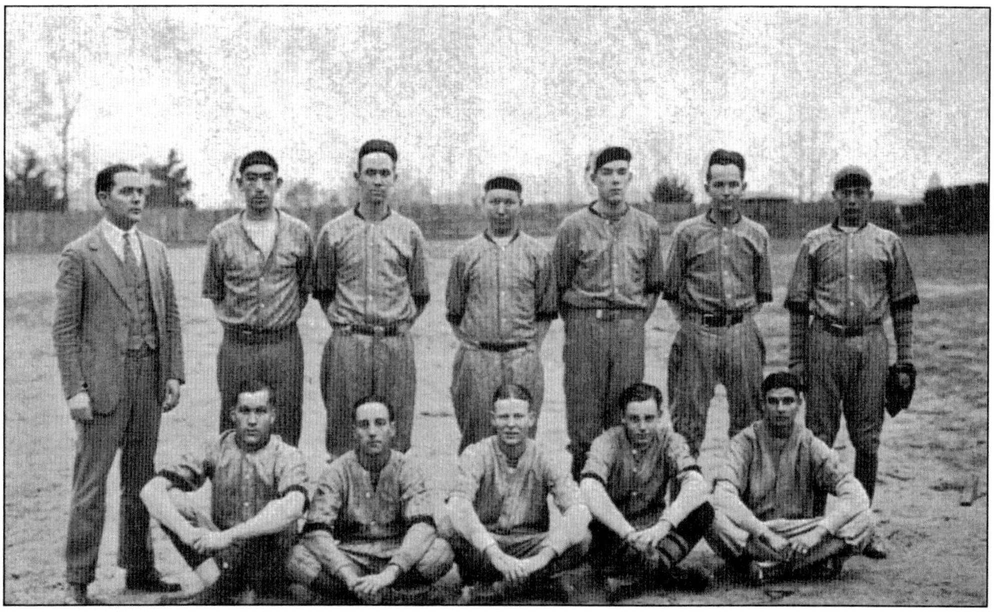

The 1927 Sheridan baseball team is shown here east of the 1913 school building. (Courtesy of Sheridan School District yearbook.)

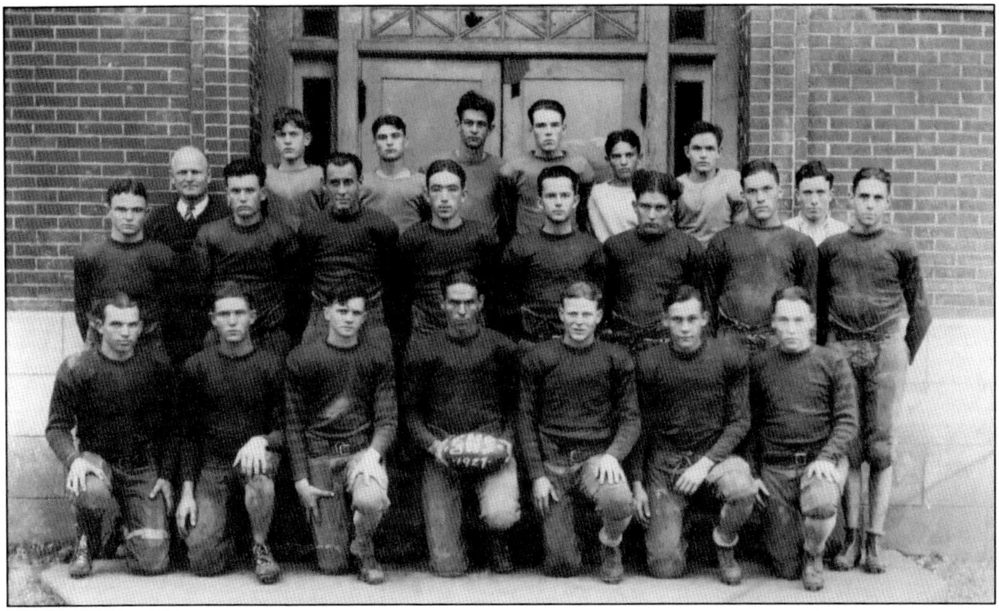

The 1927 Sheridan High School Yellowjacket football team poses for a photograph. Shown here are, from left to right, (first row) Herman Knowles, Lee Drake, Hershel Goolsby, Hollis Mathis, Macon Gean, Clyde Koon, and Red Varnell; (second row) John Cole, Thomas E. Toler Jr., Otho Morris, Glen Ashcraft, Regel Cotton, Iro Keesee, Frank Koon, and Vernon Mathis; (third row) superintendent and coach E.D. Brandon, Bud Davidson, Wallace Mitchell, Wilton "Witt" Stephens, Harry Ashcraft, Bob Sumners, Ralph Nelson, and O.C. Green, assistant coach. (Courtesy Grant County Museum.)

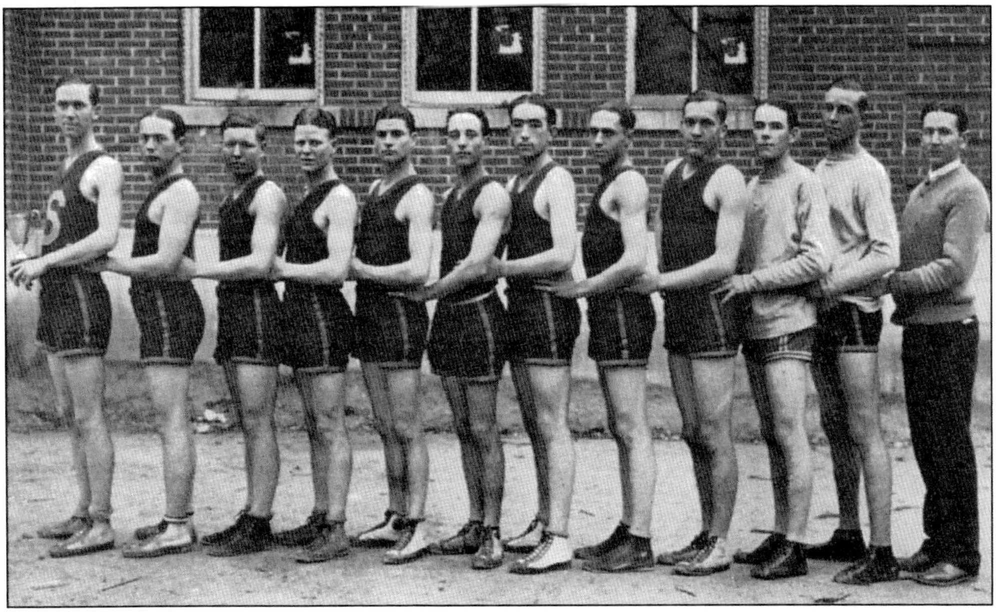

The Sheridan Senior basketball team is assembled for a photograph. Shown here are, from left to right, J. Dee Lybrand (captain), Dudley Herrington, Dennis Murdock, Macon Gean, Wallace Mitchell, Elwin Johnson, Glen Ashcraft, Vernon Mathis, Frank Koon, Harry Ashcraft, Brooks Mays, and coach Elmer Smith. (Courtesy of Sheridan School District yearbook.)

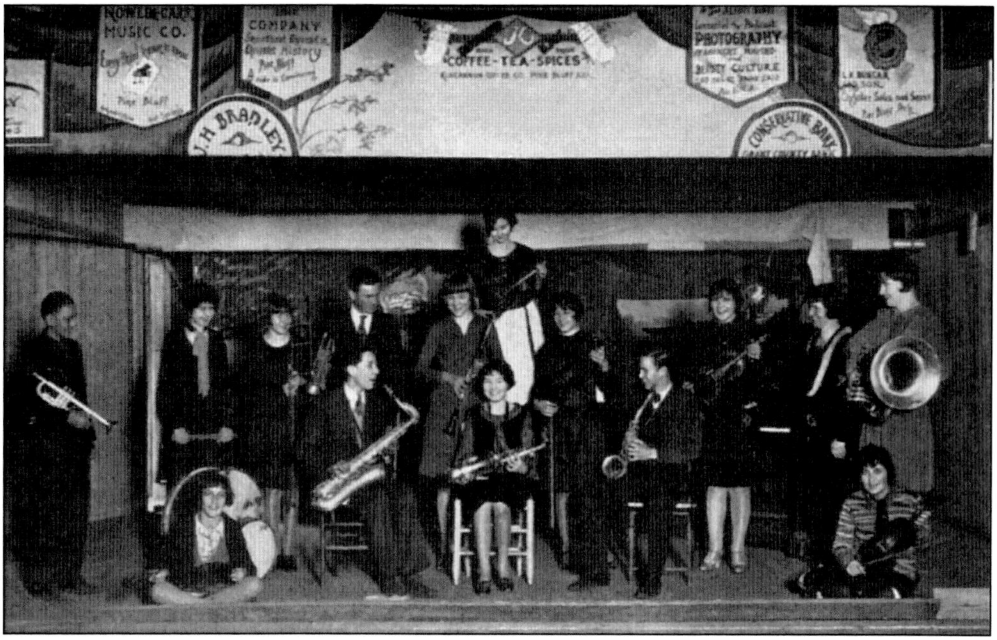


The Sheridan School Orchestra, organized by Lois Ferguson, cheerfully poses onstage in the Auditorium–Community Building. Added in 1926 were a tenor banjo by Odessa Swaim, clarinet by Retha Jones, trombone by Dee Lybrand, and ukulele by Bernice Mathews. Located a few yards northeast of the 1913 school, the auditorium had a gravel floor and was used in the summer as a canning kitchen for townspeople in need of help with preserving food. (Courtesy of Sheridan School District yearbook.)

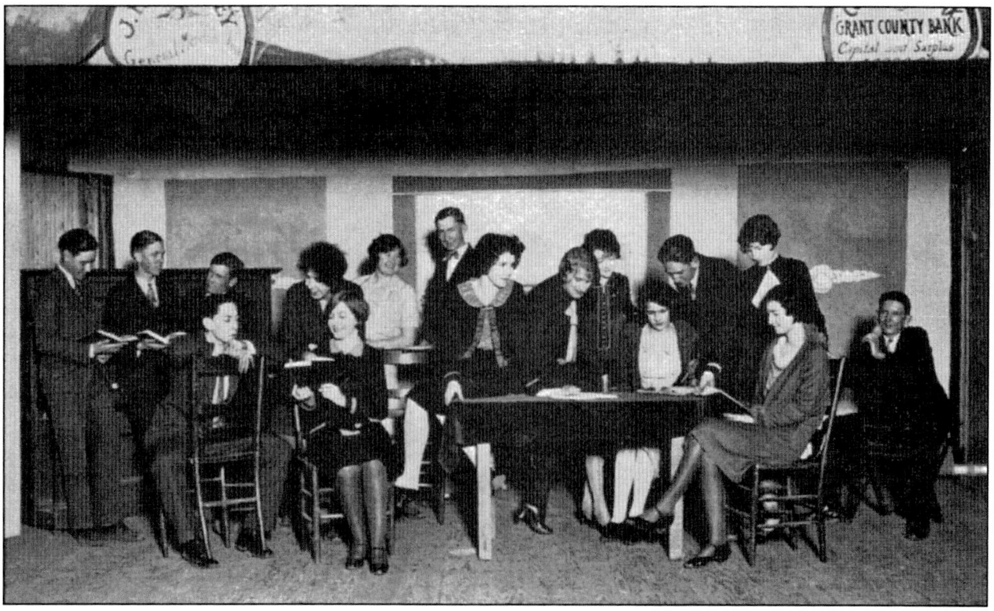


Mary Cabiness stands fourth from left behind a chair among members of Sheridan's Latin Club. The club, posing onstage in this photograph, searched for advancement in knowledge. In the 1920s, the school newspaper was called the *Booster*; it was renamed the *Tattler* in 1930. It was published semimonthly and sold for a nickel. (Courtesy of Sheridan School District yearbook.)

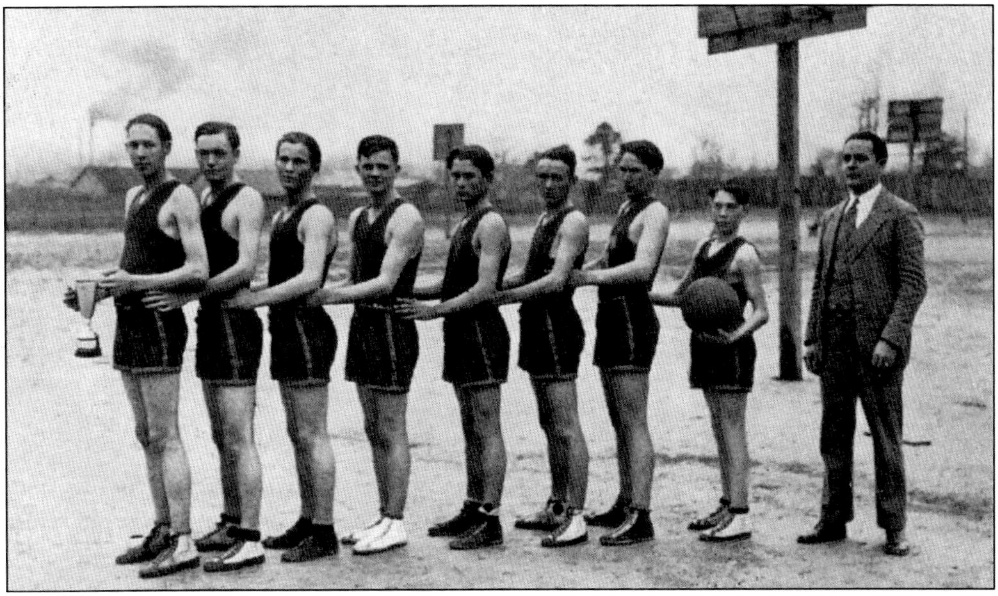

Sheridan's 1927 junior basketball team was the unbeaten regular-season Grant County champions but lost the district contest by six points to Pine Bluff. Hershell Goolsby, captain, received all-district and all-state second team honors. Shown here are, from left to right, Lewis Howard, Carroll Rushing, Byford Swiney, Hershell Goolsby, Herbert McCool, George Berry, Lee Barnes, Wilton Goolsby, and coach L.A. Francis. (Courtesy Sheridan School District yearbook.)

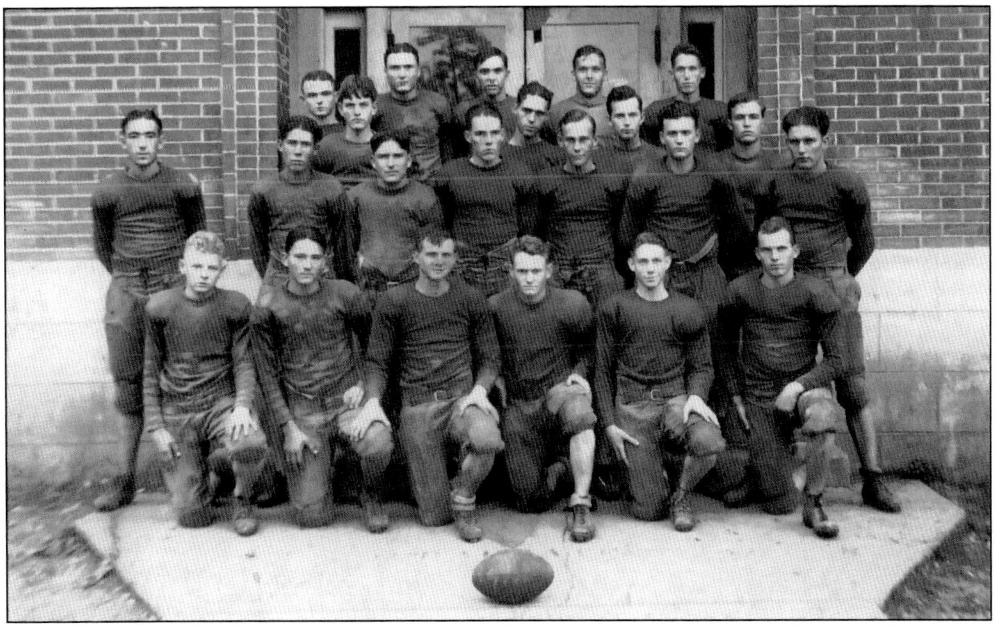

The 1928 Sheridan Yellowjacket football team poses for a photograph. Shown here are, from left to right, (first row) Jim Cole, Hershel Bradley, Hershel Goolsby, Scott Harper, Wilton Goolsby, and Herman Knowles; (second row) Glen Ashcraft, Zoda Haley, Lee Barnes, Harry Ashcraft, Clyde Koon, Tom Toler, and Iro Keesee; (third row) Bud Davidson, Vernon Mathis, Reginald Cotton, and Lealdon Smith; (fourth row) John Cole, Don Walters, Fred Jordan, Frank Koon, and Lewis Howard. (Courtesy of Doug Lloyd.)

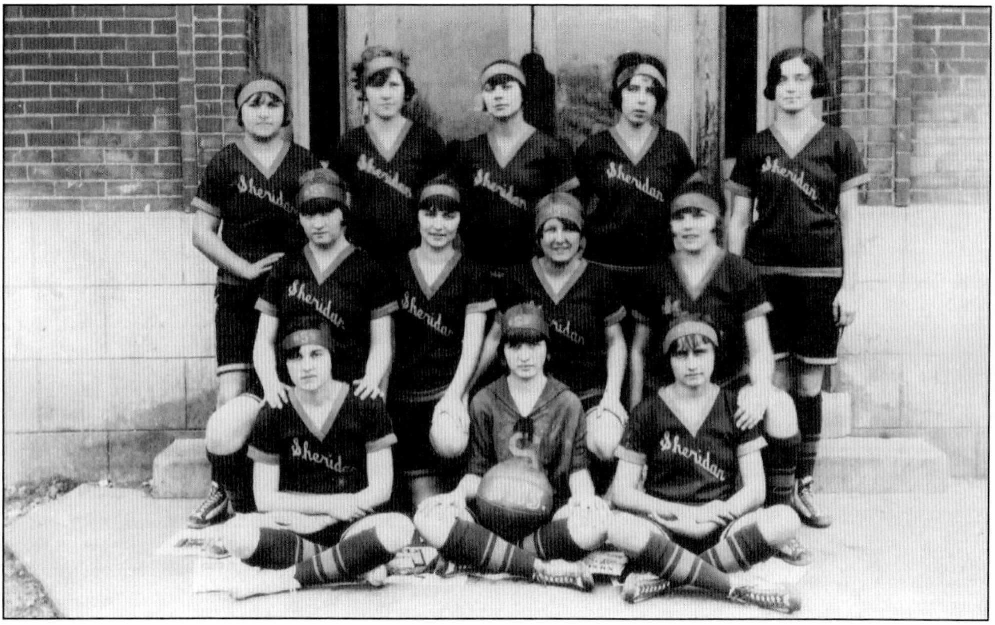


Players on the 1928 Sheridan High School girls' basketball team are, from left to right, (first row) Ruby Lee Koon, Aline McDonald, and Jamie Warlick; (second row) Pearl Pierce, Negie McCool, Mary Wilson, and Odessa Stuckey; (third row) Sibyl Bratton, Joy McDonald, Elma McMillan, Mary Cabiness, and Bertha Ballard. (Courtesy of Doug Lloyd.)

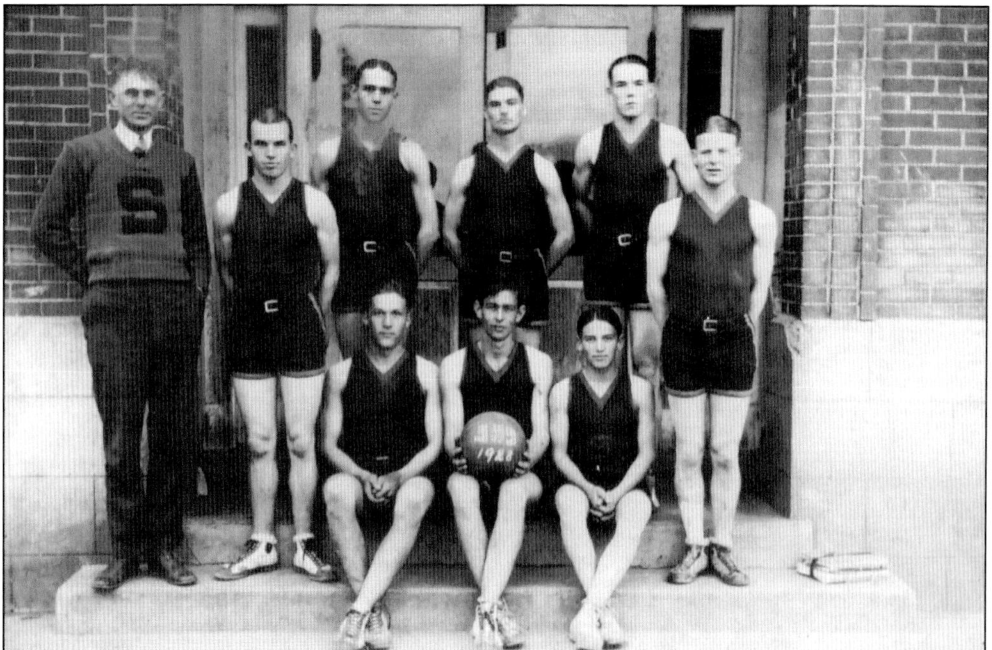

The 1928 Sheridan High basketball team poses on the front steps of the 1913 school building. Shown here are, from left to right, the following: (first row) Frank Koon, Wilton "Witt" Stephens, and Robert Sumners; (second row) coach Gus McMillan, Herman Knowles, Vernon Mathis, Wallace Mitchell, Harry Ashcraft, and Macon Gean. (Courtesy of Grant County Museum.)

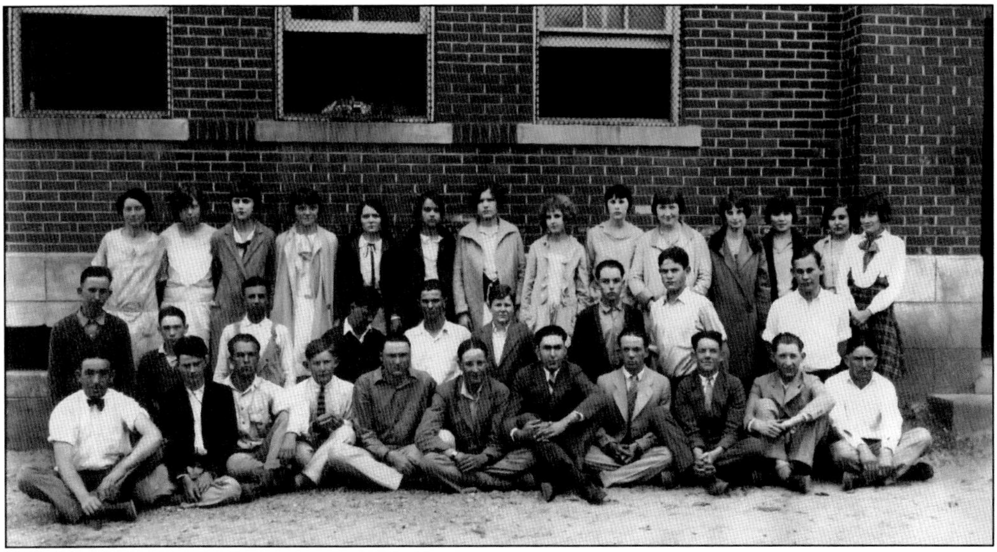

Members of the Sheridan High School freshman class of 1928 are, from left to right, (first row) Glen Ashcraft, unidentified, Lealdon Smith, Vernon West, Don Walters, Dee Jennings, Howard Clark, Milburn Mitchell, Noble Dodd, Willie Payton, and Lee Barnes; (second row) Laurence Roush, Wilton Goolsby, Fenter Ballard, Hershel Goolsby, Byford Swiney, Eldred Gean, Loran Worthen, Herman Davidson, and Clyde Koon; (third row) Alice Hope, Dovie Pinkerton, Geneva Love, Retha Jones, Veta Reid, Bernice Mathis, Madge Byrd, Leola Knowles, Lyndell Walker, Irene Byars, Bernice Mathews, Reba Stevenson, Mildred Harper, and Lena Speck. (Courtesy of Doug Lloyd.)

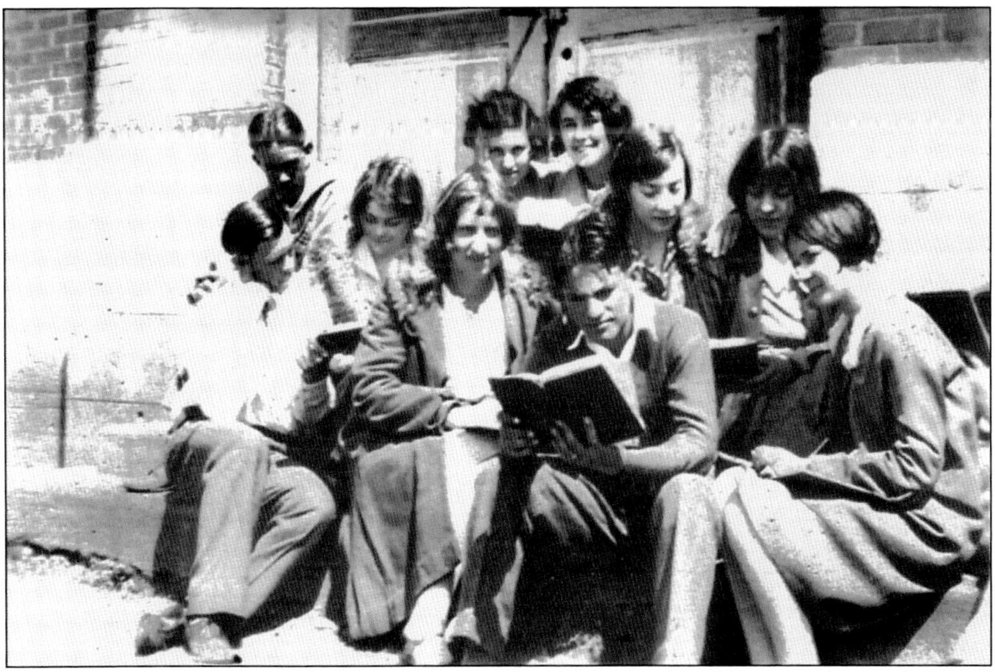

Gathered at one of the entrances to the 1913 Sheridan School building are these scholars in the 1920s. The question "Studying French?" is written on the photograph's reverse. (Courtesy of Grant County Museum.)

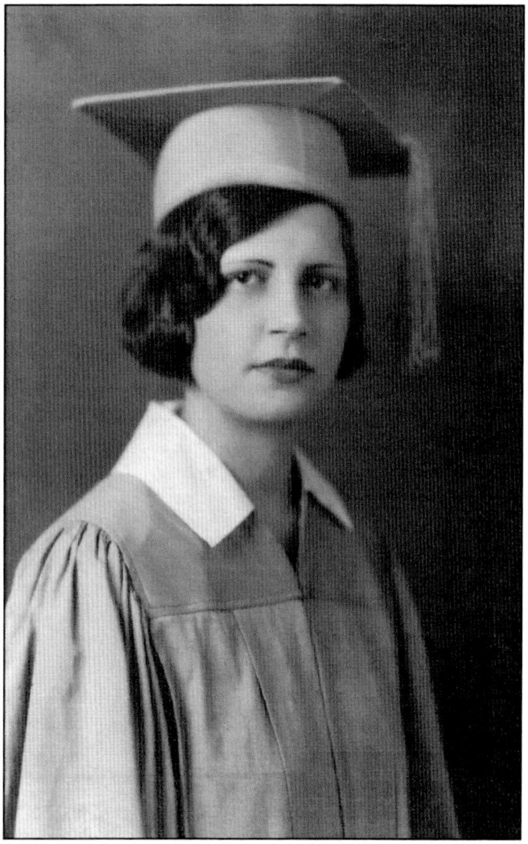

A graduate of the Missionary Baptist College on Church Street in Sheridan signed her photograph simply, "Always, Toots." (Courtesy of Grant County Museum.)

Boy Scouts of America Troop 60 of Sheridan posed for this photograph in 1944. The scouts are, from left to right, as follows: (first row) Edgar Posey, Edwin Earl Waddell, Donald Lee Key, Leon Morehead, David Jewell Jones, and George Key Mitchell; (second row) Don Blackwell, Bud Crutchfield, Bob Paxton, Johnny Douglas, Bob Tabor, Bob Bratton, and John Thomas Williams. (Courtesy of Bob Bratton.)

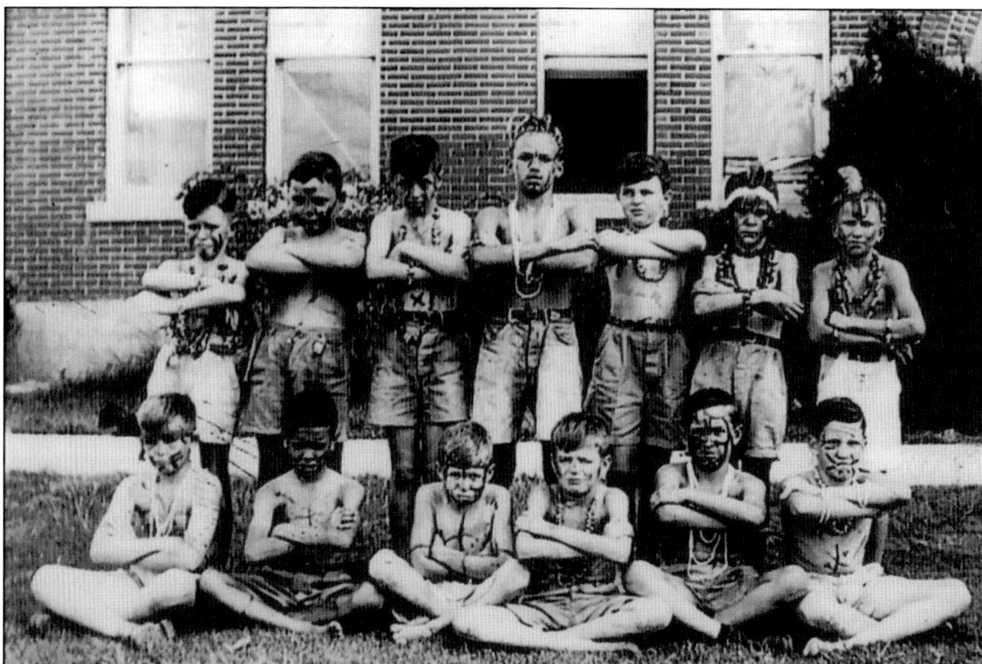

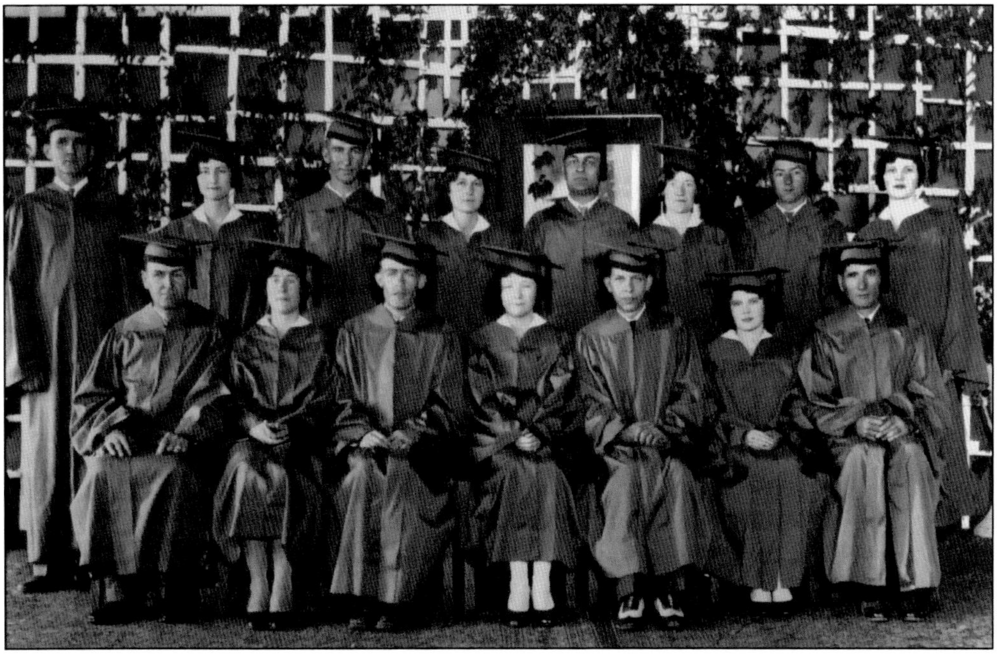

Sheridan's Missionary Baptist College existed for over a decade, offering opportunities for higher-level coursework for county students. Pictured at the 1929 graduation are, from left to right (first row) Rev. Raymond R. Parish, unidentified, John Holiman, Emogene Hope, Carroll Henslee, Willie McAllister, and Reverend Davis; (second row) Rev. Conrad N. Glover, Hazel Mathews, John Bell, Ollie Harrington, Homer Ashcraft, Ila Byers, Clinton Bradford, and Mildred McBurnett. (Courtesy of Grant County Museum.)

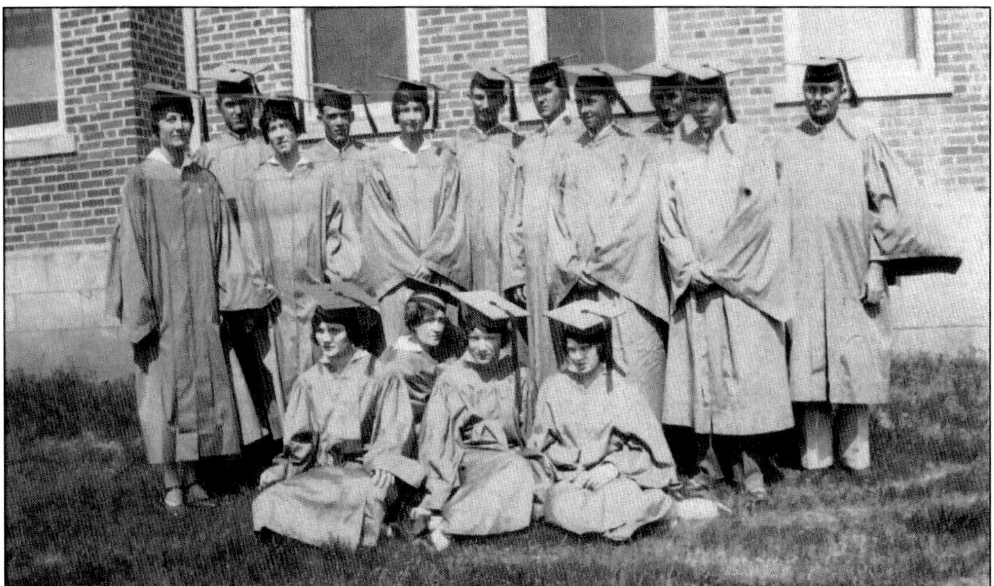

Poised to take on the world in 1931 is this group of graduates of the Missionary Baptist College in Sheridan. Had the Great Depression not restricted the college's enrollment, one can only guess how large it could have grown. (Courtesy of Grant County Museum.)

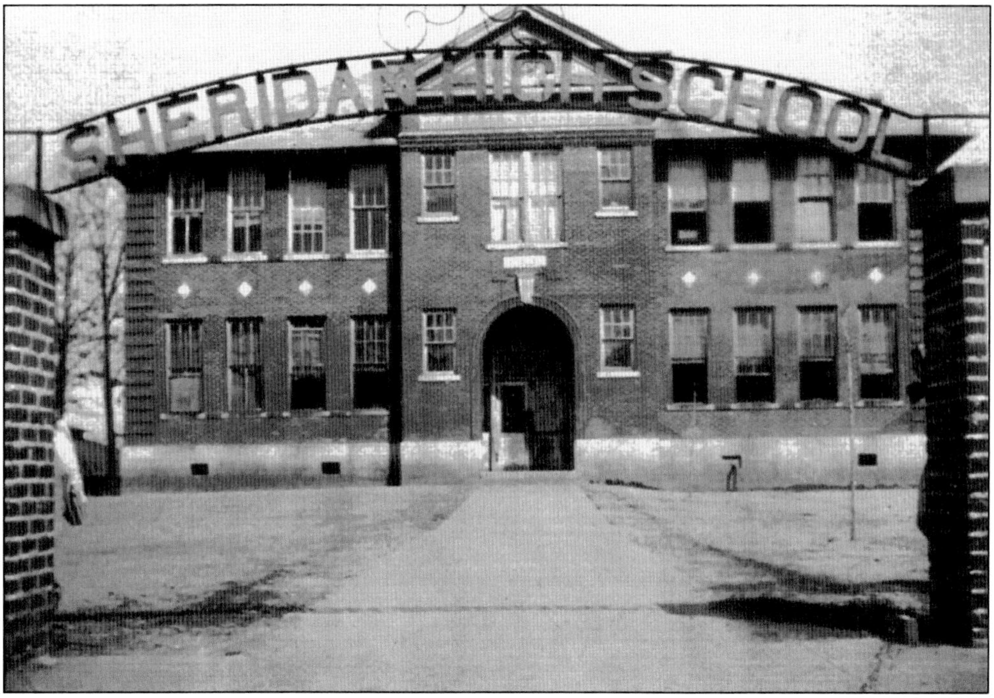

The Sheridan High School sign, pictured here in 1930, stood for five decades before being destroyed by a traffic accident. (Courtesy of Grant County Museum.)

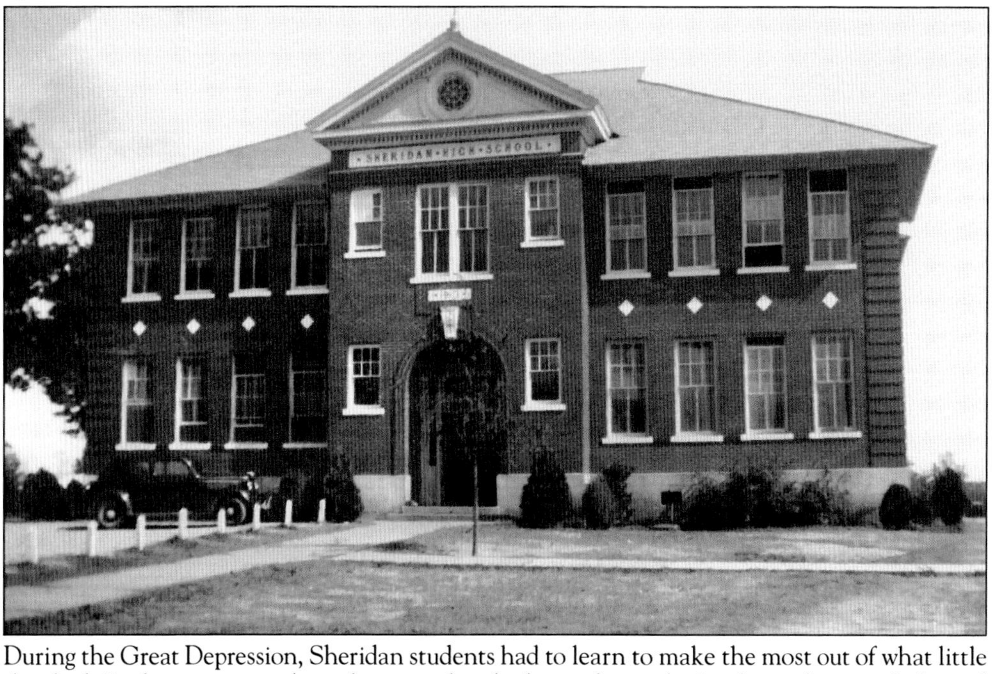

During the Great Depression, Sheridan students had to learn to make the most out of what little they had. Students were taught an honor code, which saved a teacher's salary otherwise dedicated to keeping study hall. Lunch was brought from home, often a biscuit with only turnip greens inside, and perhaps a sweet potato. Teachers heated milk on a stove for the younger students. (Courtesy of Grant County Museum.)

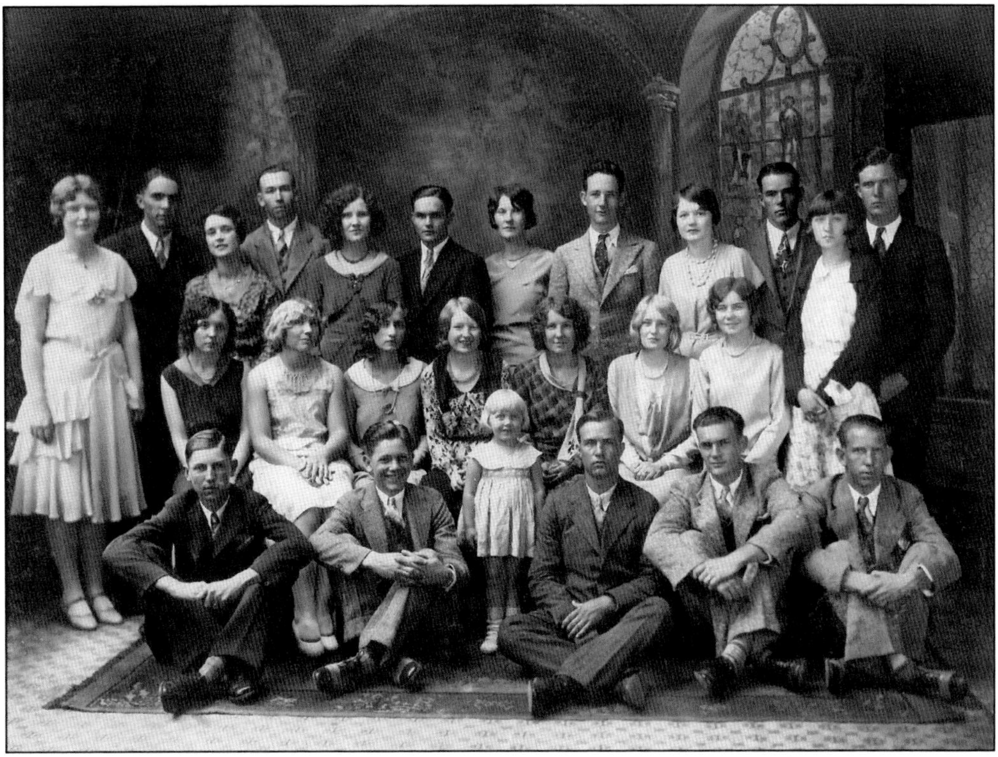

The 1930 Sheridan High School senior class poses for a photograph. Shown here are, from left to right, (first row) Ottis Ashcraft, Frank Springer, mascot Francis Gwin Williams, Lealdon Smith, Thomas E. Toler Jr., and Mack Williams; (second row) Vera Murdock, Newell Nelson, Mary Wilson Reaves, Clyde Newell Walker, Glennie Mae Mingea Springer, and two unidentified; (third row) Dora Belle Pinkerton Chronister (class president), Fred Jordan, Valerie Henderson Jordan, Hartwell Davis, Joy McDonald Nutt, Ralph Nelson, Helen Drake Barnes, Adrian Johnson, Lexa Jones Buie, Harry Ashcraft, Rebecca Monroe Bell, and Lee Drake. Not pictured are Nellie Perry Bruce, Edna Murry Tabor, Leon Reid, Guy Mayfield, Willie Pinkerton, and Herman Knowles. (Courtesy of Benny Barnes.)

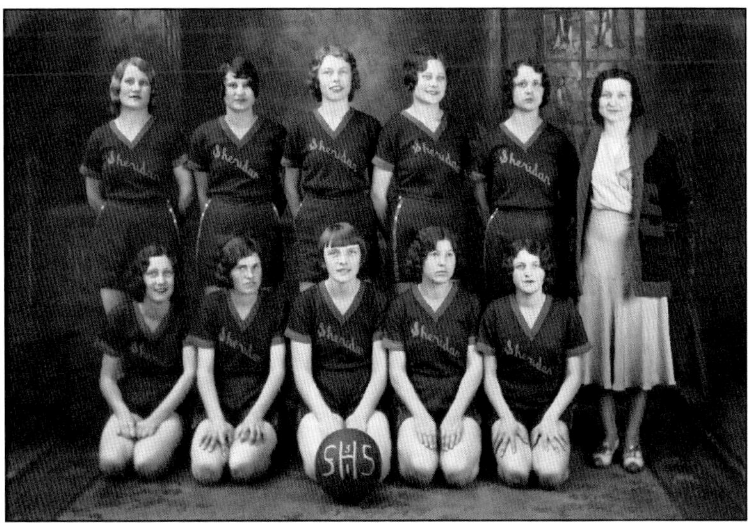


Members of the 1931 Sheridan girls' basketball team are, from left to right, (first row) Gladys Ashcraft, Louise Harper, Retha Jones, Joe Wilson, and Floy Vanlandingham; (second row) Ethel Hulse, Lillian Reese, Dovie Pinkerton, Floy Wright, Bernice Mathis, and Irene Berry (coach). (Courtesy of Doug Lloyd.)

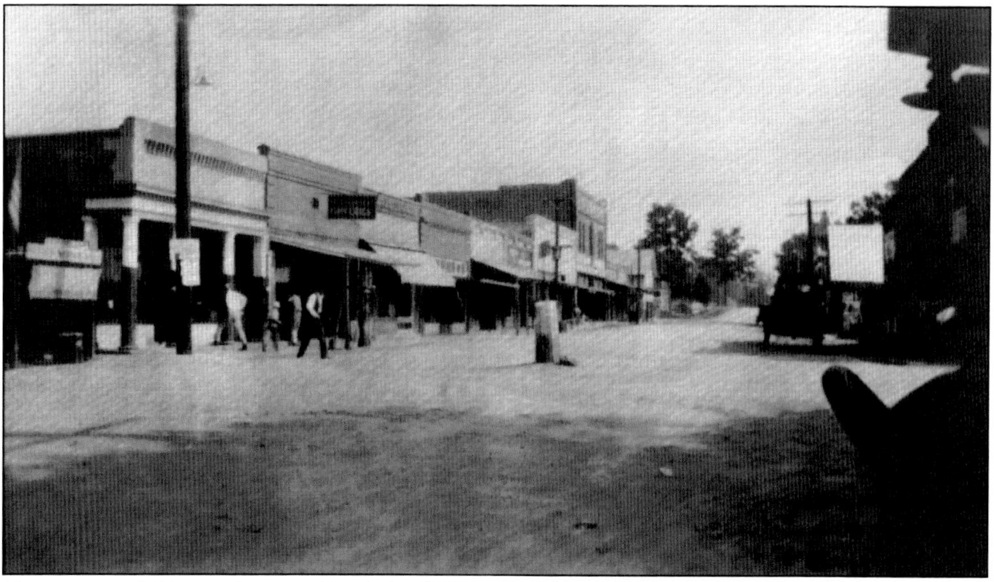

Customers of Grant County Bank (at left with columns) begin their day at the corner of Oak and Center Streets in Sheridan. A mid-intersection post announced town events. The Presbyterian church tower is barely visible at right above the parked truck. (Courtesy of Grant County Museum.)

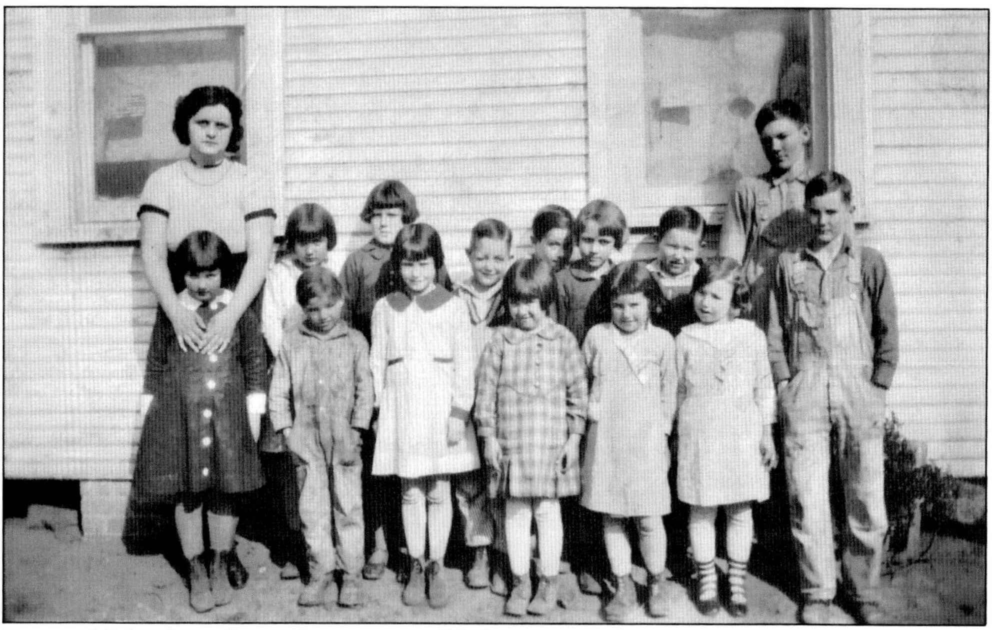

Dogwood School stood near the junction of Highways 46 and 291 in southwestern Grant County. Grades two through four, pictured here in 1935, are, from left to right, (first row) Ida Lee Duncan, unidentified, Eva Lou Threet, Aleen Moore, Helen Crutchfield, Louise Threet, ? Macmon; (second row) teacher Eva Gill, Dorothy Crutchfield, Mary Lewis Threet, Henry Harrington, Arlee Smith, unidentified, Charles Threet, ? Macmon. (Courtesy of Grant County Museum.)

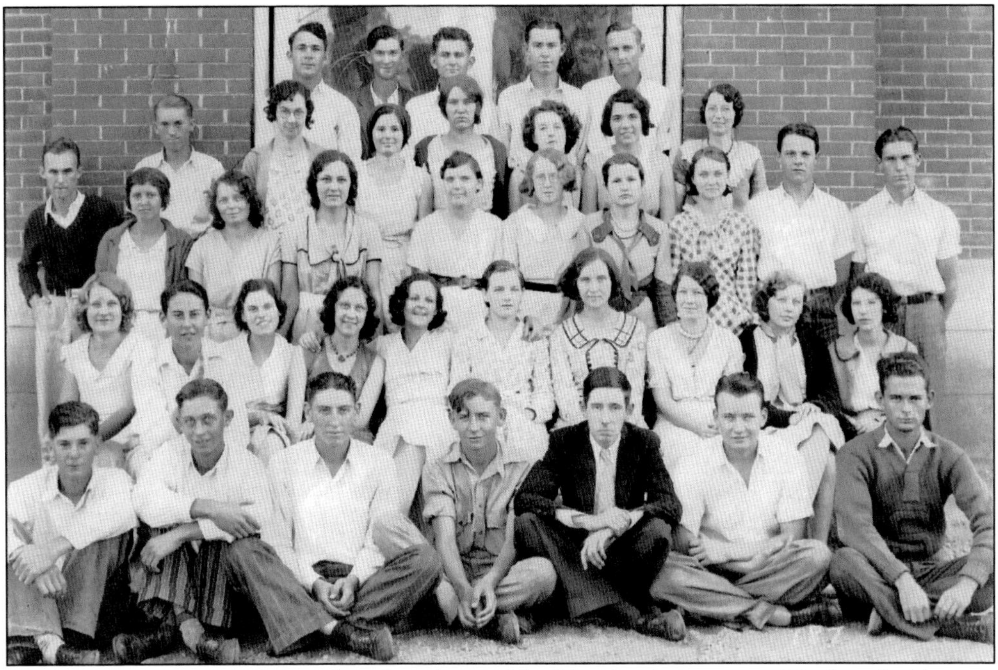

Sheridan's 1933 senior class gathered at the 1913 building's side entrance. Unidentified by row, graduates were Joe DuVall, president; Basyl Gentry, vice president; Bruce Gartman, secretary; Louise Harper, treasurer; Dale Bratton, reporter; and Gladys Ashcraft, Mary Ellen Kelly, Ethel Mae Nelson, Bob McKenzie, Vivian Batchelor, Clinton M. Atchley, Charlene Mathews, Leonard Reeves, Joelma Wilson, Charles W. Reid, Geraldine Williams, Aubrey H. Bradford, Henrietta Everett, Gayle Edward Gunn, Frances Lorene Key, Herbert McCool, Emma Grace Taylor, Jack E. Reynolds, Mary Evelyn Reese, Basil McGarity, Margaret McDonald, Ethel M. Hulse, William H. Jackson, Maxine Threet, Kirby Stevenson, Lorine Goolsby, Robert Edward Lee, Daisy Van Gentry, Donald Arthur Meek, Leta Mae Hood, Bruce Gwin, Lyndell S. Hamilton, Clark Salley, Ennis E. Greene, S.F. Fielding, Gertrude Smith, Lucille Stuckey, and Howard Shoptaw. Mrs. O.W. Hope was sponsor and Alton H. Thomas superintendent. (Courtesy of Grant County Museum.)

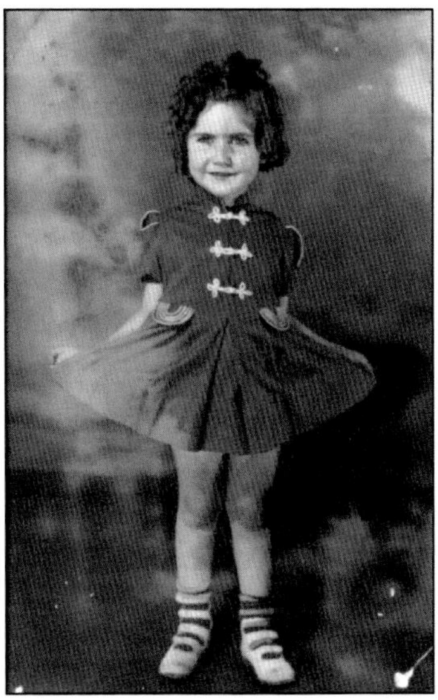

Mascot of the Sheridan senior class of 1936, four-year-old Betty Lou Thornton was known for her Shirley Temple looks and talent. This photograph was given to Alice Hope, senior class sponsor. Betty Lou's father, Raymond H. Thornton, served as Grant County school superintendent, and her mother, Wilma Stephens Thornton, taught in Grant County for over four decades. (Courtesy of Grant County Museum.)

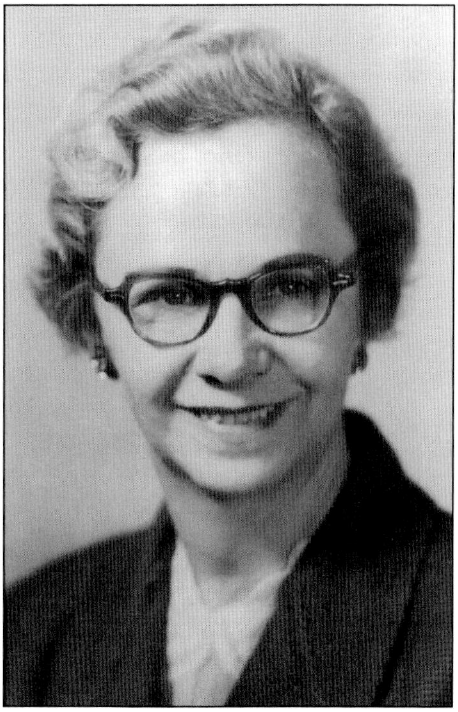

Minnie B. Humphrey Clark, the first female sheriff in Arkansas, was appointed by Gov. Carl Bailey in 1937 upon the death of her husband, Sheriff Paul J. Clark Sr., with four months remaining in his term. Born in Bauxite in 1907, Minnie Clark was elected sheriff in 1938. She was elected Grant County tax assessor in 1940 and county treasurer in 1944. Before her political career, she taught in a one-room county school for one year. She lived until 1993 and was buried in Sheridan Cemetery. (Courtesy of Grant County Museum.)

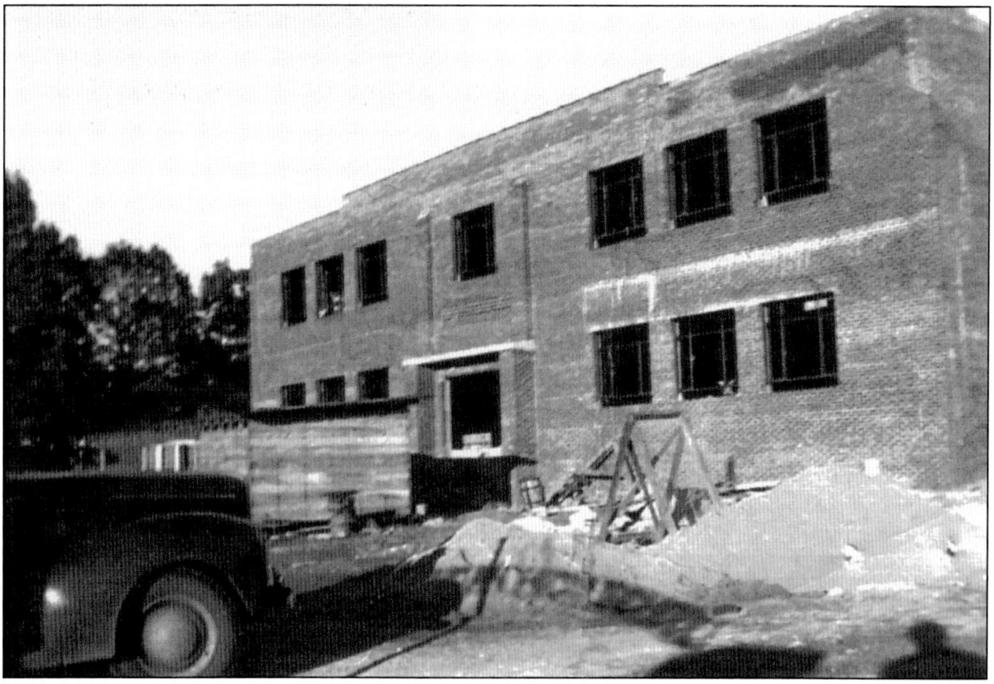

Kelly Williams Gymnasium, constructed in 1947, was touted at the time as the most modern in the state. It was named to honor Dr. O.R. Kelly, longtime supporter of education and member of the school board, and the Williams brothers, owners of the nearby sawmill, who donated large amounts of lumber for the project. Classrooms were upstairs along the catwalk, and the building housed the school library. (Courtesy of Grant County Museum.)

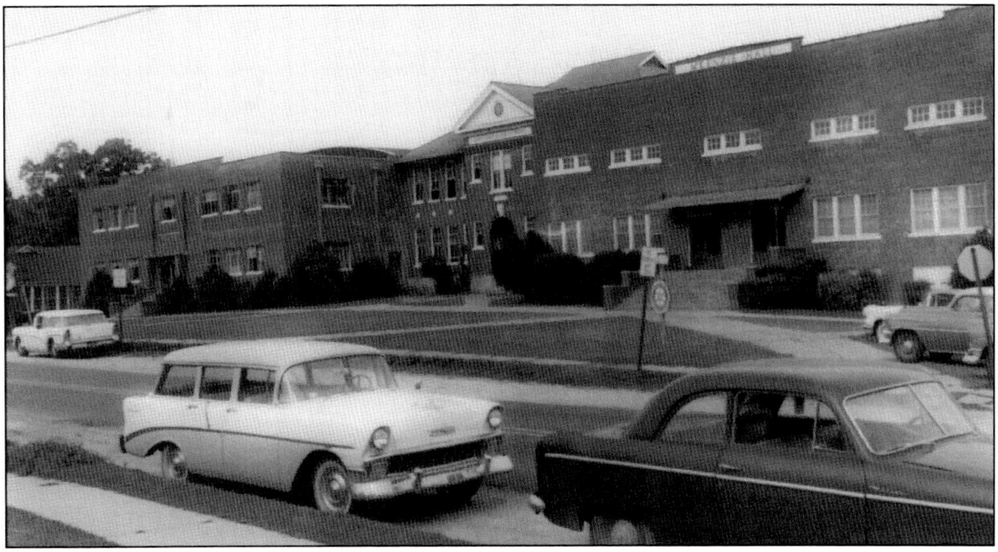

The decade of the 1950s was the last to see the entire high school housed at the 1913 building complex, with the gymnasium (1947) attached to the north and McKenzie Hall (1942) to the south. Soon after the first windowless high school building in Arkansas was constructed in 1963 just to its north, rapid growth called for a larger high school complex to be built seven blocks west, in 1970. (Courtesy of Grant County Museum.)

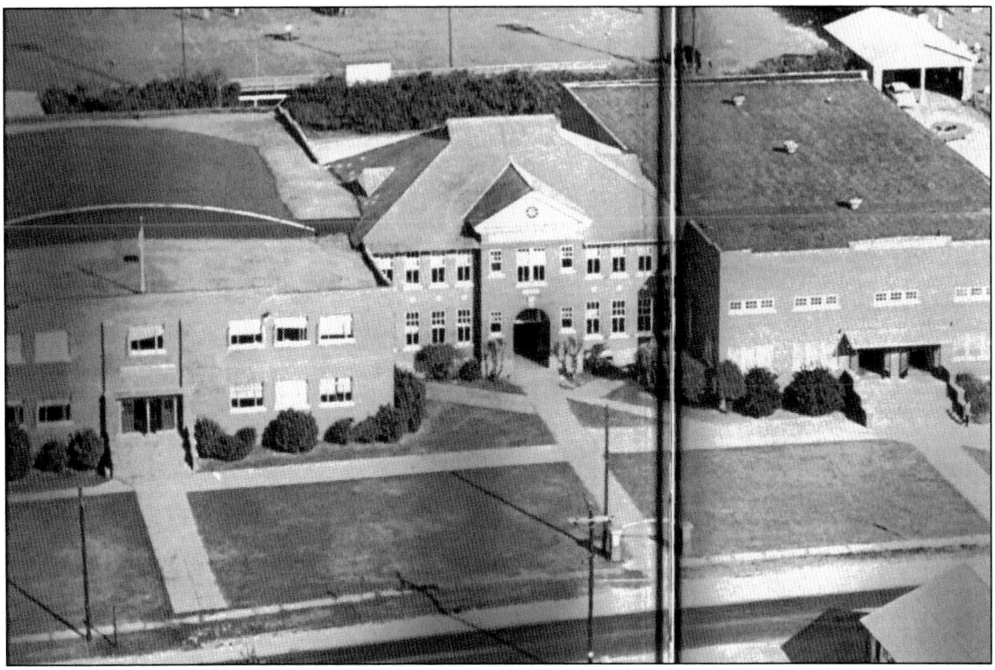

In this photograph, football practice occurs on the original field, which served until the 1980s. Named after A.R. McKenzie, Sheridan superintendent from 1933 to 1962, McKenzie Hall, at right, provided seating for 1,000 and contained the first cafeteria downstairs. State and federal officials spoke from its stage. In the 1970s, country music star Jerry Lee Lewis performed. In the 1980s, first lady of Arkansas Hillary Rodham Clinton held an education town hall meeting there. (Courtesy of Sheridan School District yearbook.)

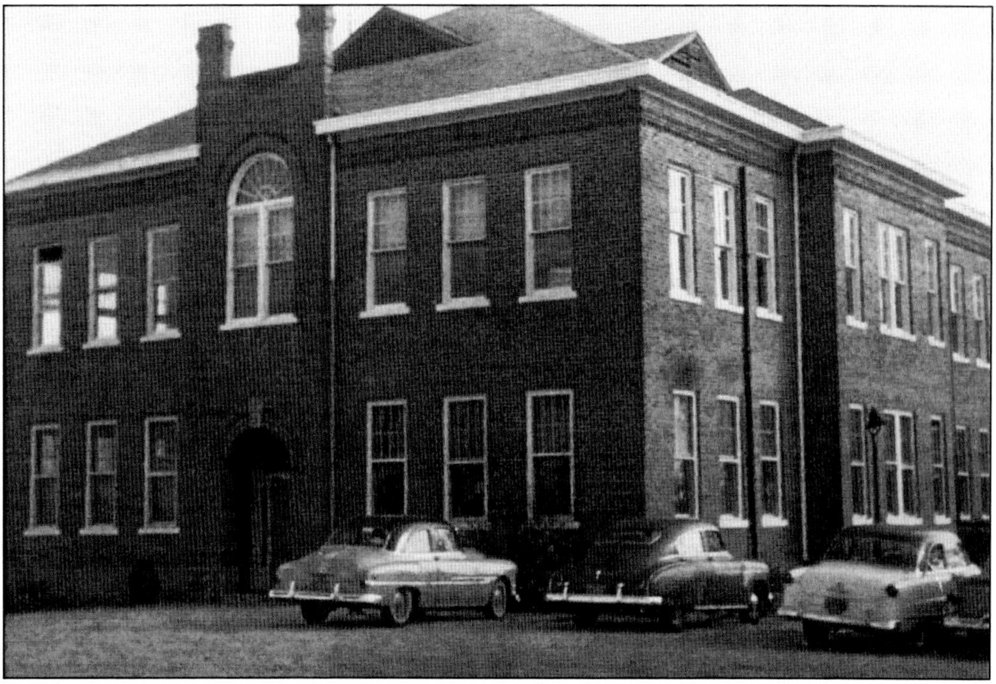

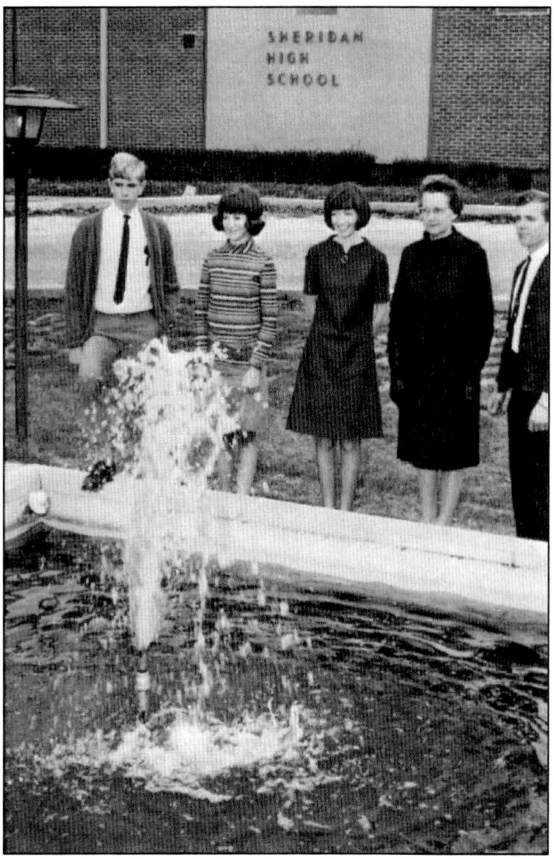


In 1950, there were 51 teachers in the Sheridan School District, which then included the former Missionary Baptist College building and East End campuses for elementary students. This college building, patterned after the 1913 building, was built in 1921. The Sheridan District acquired the property in the 1930s. The building burned in 1956 and was replaced by an elementary campus, which occupied the site until 2004. (Courtesy of Grant County Museum.)

Soon after the 1963 Sheridan High School building was completed, the student council built a fountain dedicated to sponsor Wilma Stephens Thornton. The project, with Arkla Gas lighting and colored floodlights, was intended to lift the spirits of the student body. Shown here are, from left to right, Tim Armstrong, Marie Crawford, Kathy Grace, Thornton, and C.W. St. John. (Courtesy of Sheridan School District yearbook.)

# *Five*

# 20th-Century Memories

As the 20th century dawned in Sheridan and Grant County, progress was slow. There was no electricity, no indoor plumbing, no automobiles—just horses, wagons, or walking. Electricity would not arrive until the 1940s for most county residents. Most families had to build their own houses, deliver their own babies, use home remedies when ill, hand-dig their own wells, draw their own water, grow their own food, bathe once a week or so, use coal oil lamps, make their own amusements, use sling blades to cut grass, slaughter cows and hogs, go fishing and hunting for food, not sport, wear homemade clothes, occasionally dodge a rabid animal, and try to survive the best they could.

The pleasures of living included hearing calls of a whippoorwill or bobwhite quail at dusk, the "Lord God" ivory-billed woodpecker making its memorable sound, a panther screaming, washtubs filled with huckleberries and blackberries picked in clearings, and bream and perch making their springtime run while catfish were caught on bank hooks in the Saline. Mothers would serve caramel cake, fried apple pies, or corn bread cooked in real iron skillets on a wood stove. Families and their welcomed guests would sit on a porch shelling peas and dodging horse flies while hoping for a cool breeze. A crowd would gather to listen when battery-powered radios arrived, bringing for the first time easy yet distracting contact from the outside world. During and after the Great Depression, breadwinners started to migrate seasonally to California and Oregon for timber jobs. Some stayed there, and others returned yearly, spending hard-earned money on transportation. World War II got everybody's attention, whether they liked it or not. Headlines in the local newspaper captured a somber mood: "52 men will be inducted next week," "Cpl. Max McClellan dies in N. Africa," "Cedar Branch Sailor Loses Life in Pacific," "Sgt. Doby Barnes Awarded Silver Star," "Sheridan Senior Curtis Clark honored as he leaves to enter Air Corps," "S-Sgt. Theo Warlick and T. Sgt. Minor F. Green Receive Distinguished Flying Cross." Television in the 1950s rapidly changed local culture. Passing years brought amazing labor-saving devices, and by century's end, unbelievable technology elevated standards of living as Sheridan and Grant County steadily joined the "modern" world.

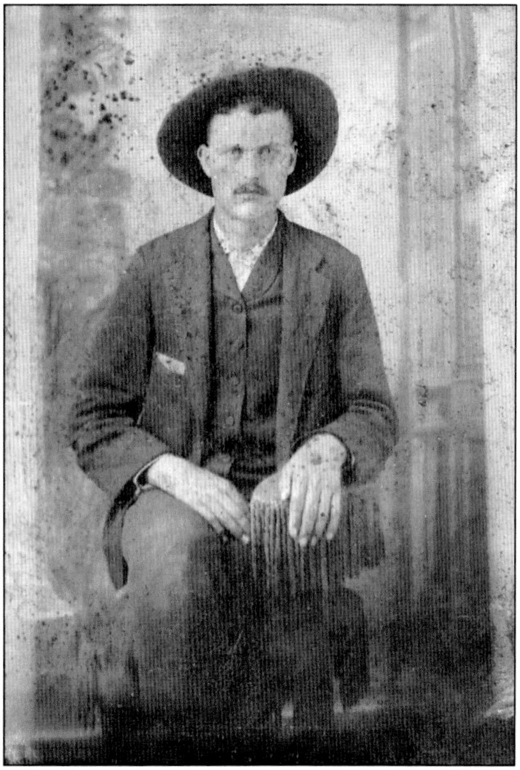

Tom Ballard, son of Lot Ballard, married Dora Fenter and lived in the Fenter community north of Poyen. (Courtesy of W.H. Byrd Jr. family.)

Abraham Tull, an early 1800s settler, found and transplanted an apple tree taken from an abandoned Quapaw farm in what became Grant County. The taste of its dried apple slices became popular, and Tull apple trees survive. (Courtesy of the author.)

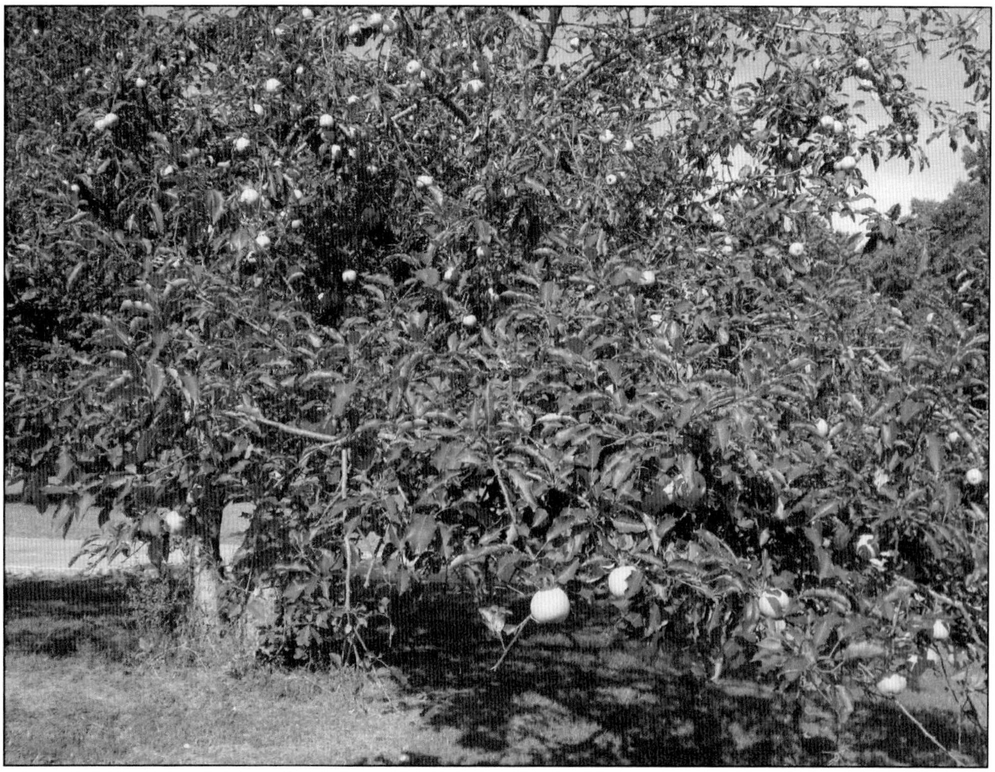

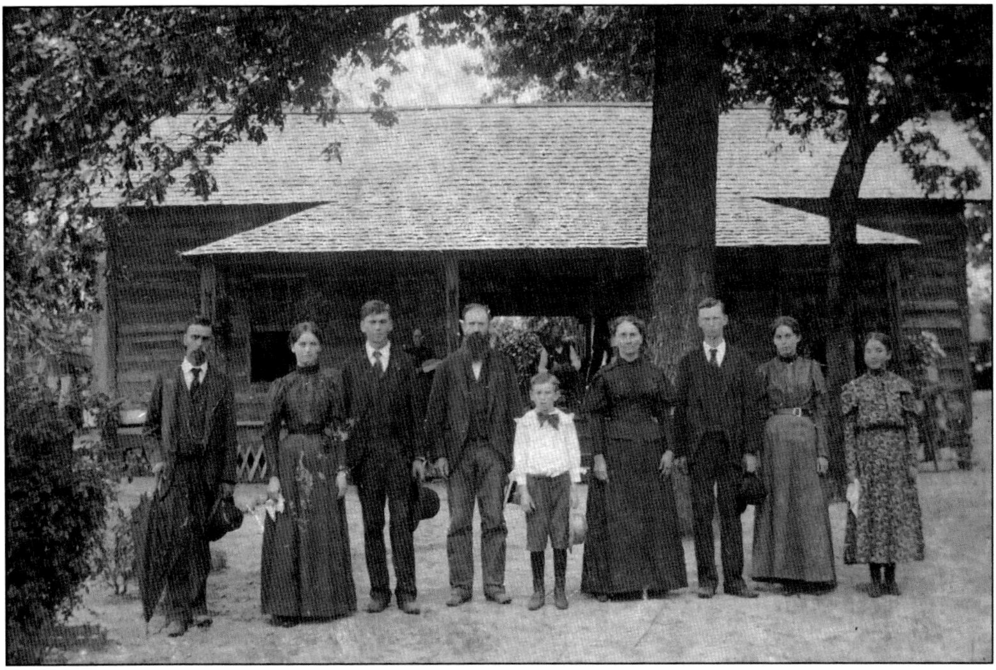

The Harrison family poses in front of its dogtrot house at Leola. A granddaughter of Dr. John W. Harrison, Mana Harrison, is at far right. (Courtesy of Grant County Museum.)

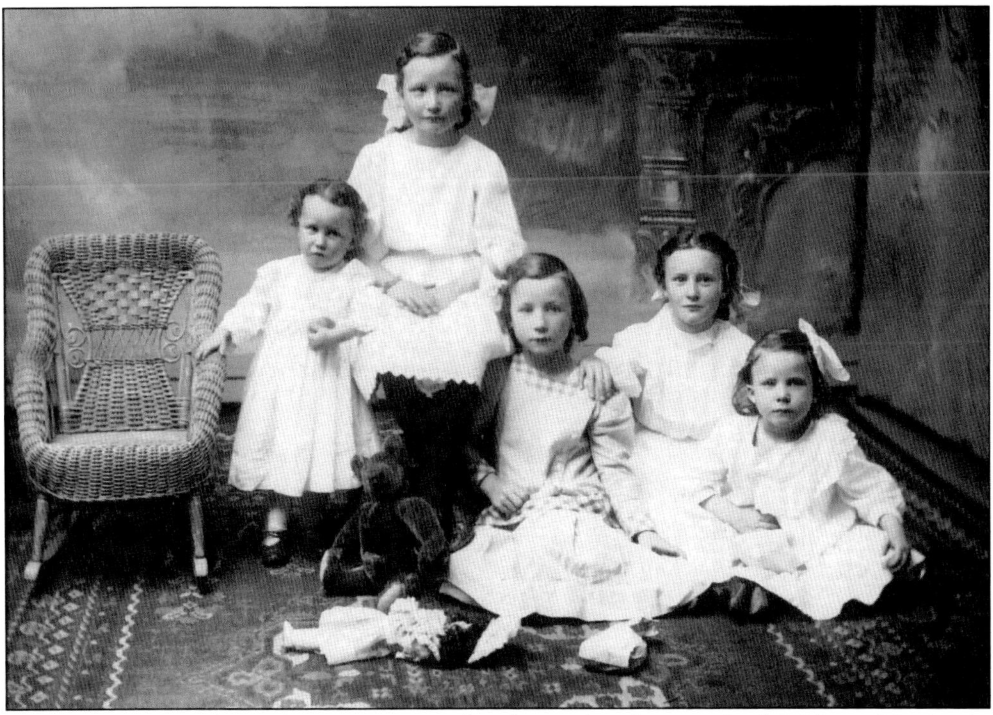

Among the papers of Dr. John W. Harrison of Leola was this photograph of children of the family, living a quality of life at the beginning of the 20th century that few in the county could enjoy. (Courtesy of Grant County Museum.)

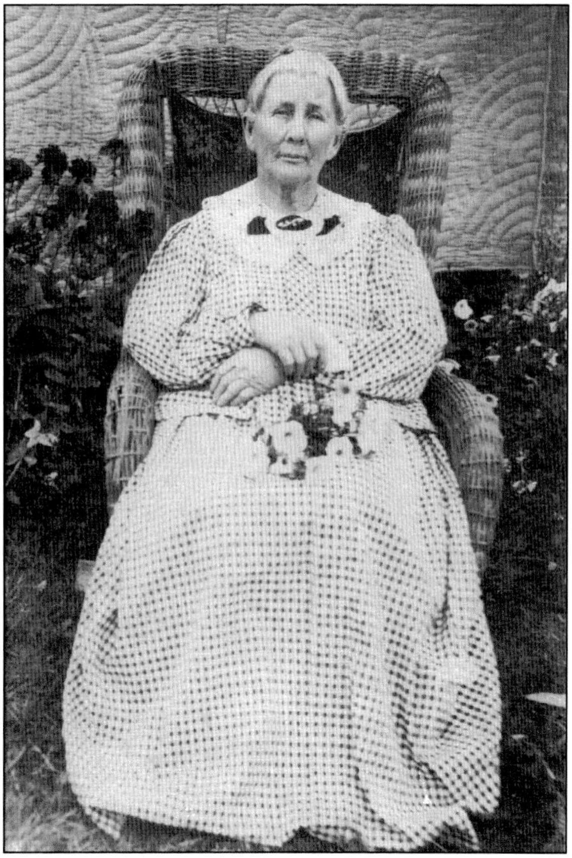

In 1860, Ann Anderson Bratton, age 13, arrived with her family to settle west of Crossroads. Her father, Isaac Anderson, born in 1795 in Rockbridge County, Virginia, said he "wanted to pioneer one more time" at age 65. Her mother, Julia Ann Schultz Foute Anderson, was born near Baltimore, Maryland. On April 30, 1864, the Union ambulance wagons, refugees, and sutlers traveled east toward Pine Bluff, passing in front of their home in what was then southern Saline County. (Courtesy of the author.)

Long before power tools made cutting timber easier, mule and horse teams were the necessary method of getting timber to market, as is evident in this Leola photograph labeled "Uncle Jack Harrison." (Courtesy of Grant County Museum.)

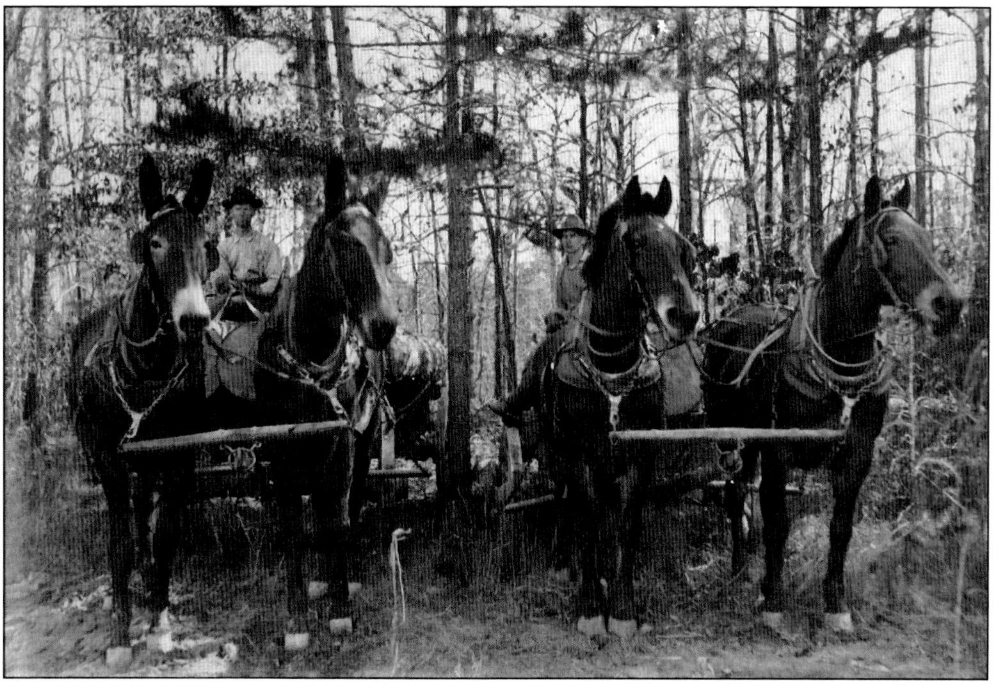

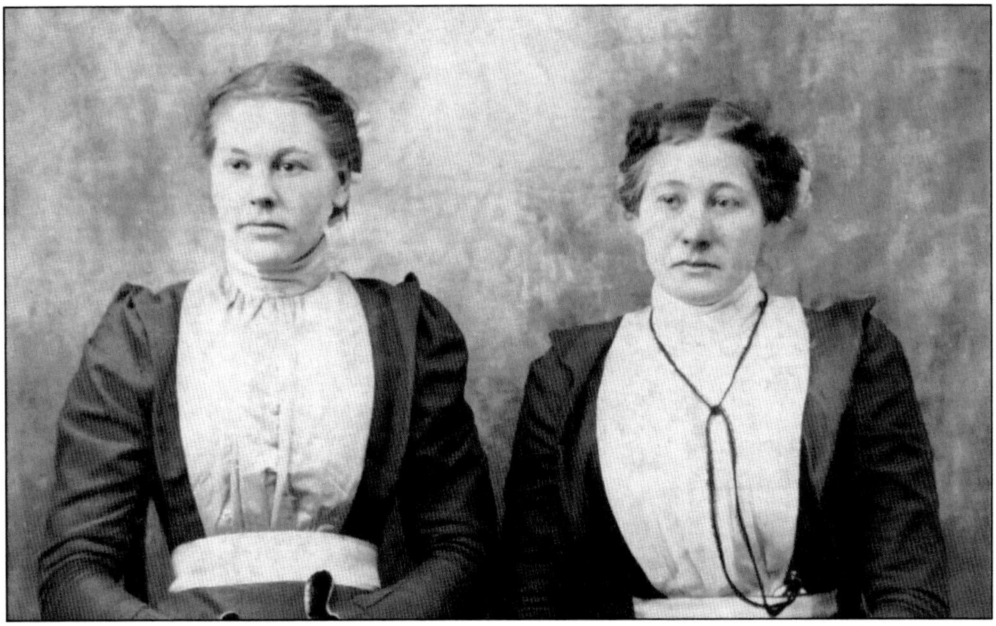

Alice May "Lizzie" Bratton, left, and her sister Laura Ellen Bratton are captured in their finest on January 4, 1901. They were two of six girls born to Thomas Robert Bratton and Ann Anderson Bratton. Lizzie, born in 1880, would always claim southern Grant County as her home. Laura, born in 1879, would eventually move to Jonesboro with her family. (Courtesy of the author.)

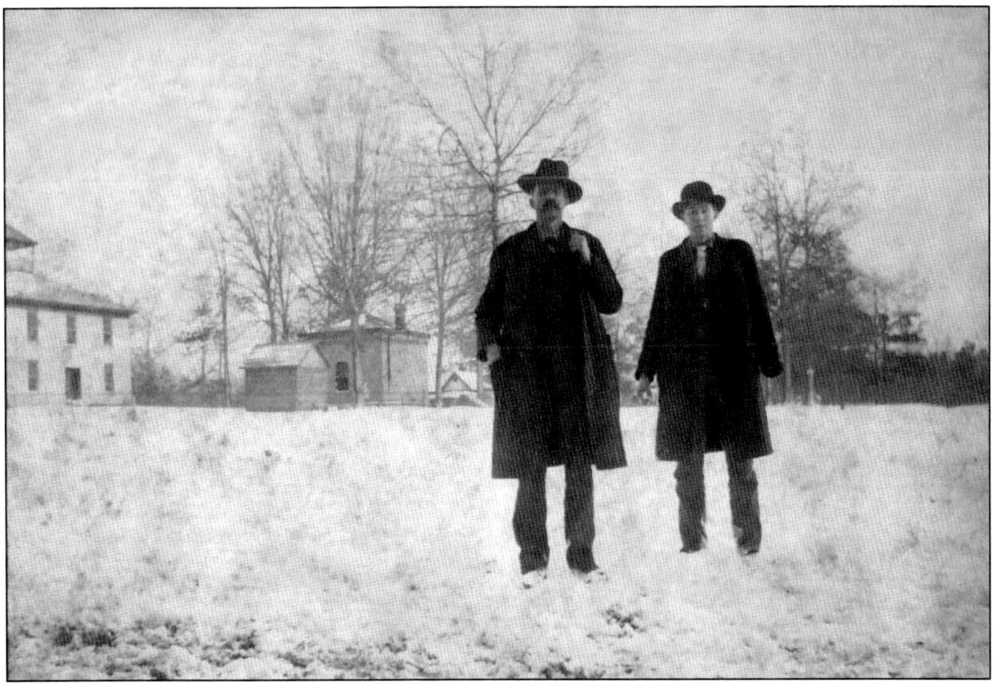

Pictured on a snowy day before 1909, with the county jail and second Grant County Courthouse in the background, are Isaac McClellan (left) and Juff Nall, who had lost his arm in a sawmill accident. As a prank, Juff's brother Tye was once locked inside the brick jail. He promptly kicked an escape through the bricks, remarking, "I don't have time for this." (Courtesy of Doug Lloyd.)

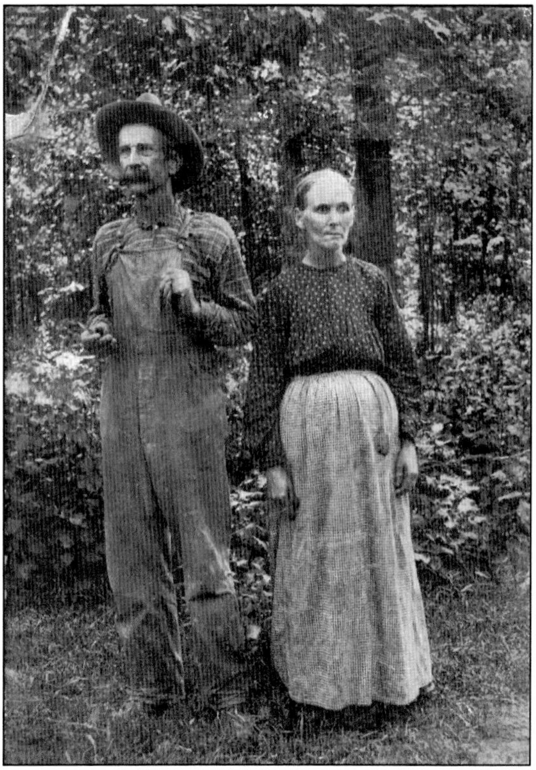

Ben Wilkinson (born September 1850) and his wife, Augusta Wilson Wilkinson (born April 1852), pose in their golden years. Ben was named after his grandfather Benjamin, from South Carolina. He was the son of Dr. Hugh Montgomery Wilkinson, a Civil War doctor and Baptist minister, and Elizabeth Patrick Wilkinson. "Gusty" was the daughter of Bradberry Wilson, who settled west of Crossroads in 1860. A descendant, Ola Hendon, operated a popular store on Center Street in Sheridan. (Courtesy of the author.)

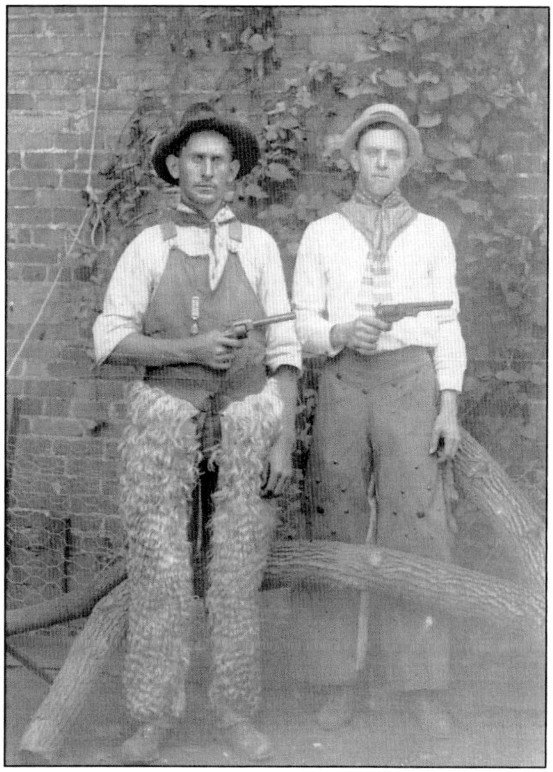

Two unidentified cowboys pose for the camera in the early 1900s. This photograph became part of the Lizzie Allen picture collection. (Courtesy of Hazel Wilson.)

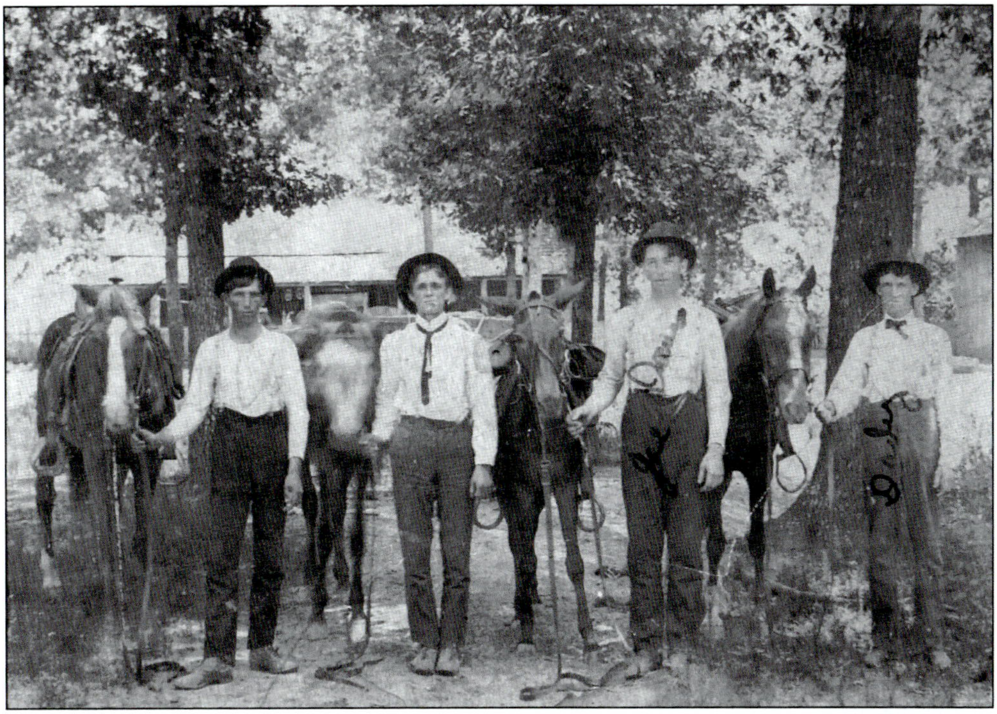

Richard Allen of the Shiloh community stands at far right, next to Joe Davis and their horses in a photograph taken in the early 1900s. Allen grew cotton, farmed, worked with timber, and made southern Grant County his lifetime home. The two at left are unidentified. (Courtesy of Hazel Wilson.)

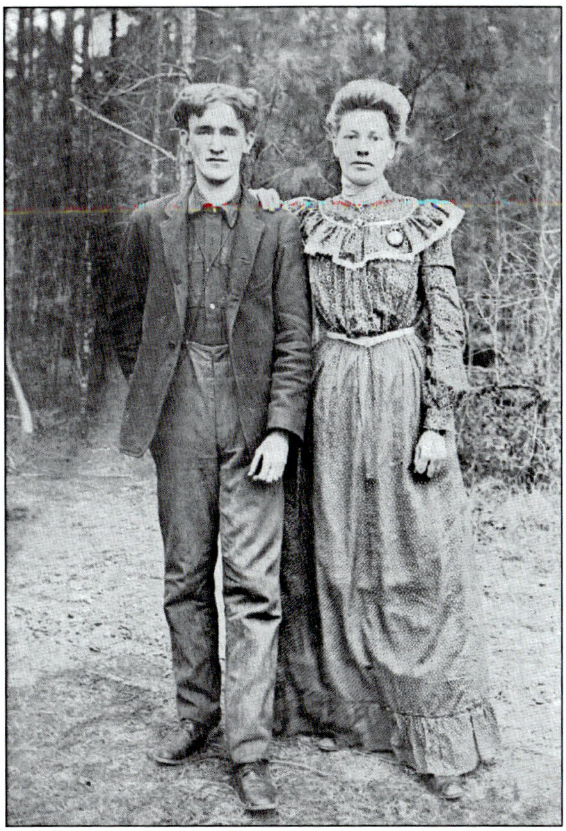

Near their wedding day in 1904, Rich and Lizzie Bratton Allen of the Shiloh community pose for a photograph. (Courtesy of the author.)

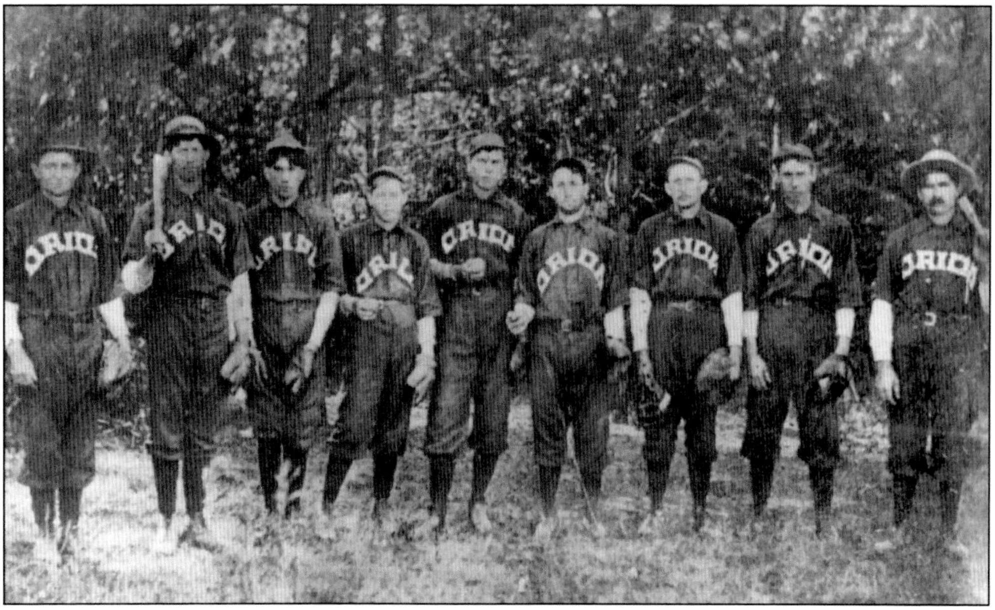

The Orion baseball team poses for a photograph in 1906. The players are, from left to right, Walter Murdock, Pat Batchelor, John Batchelor, Jody Murdock, Louis Reynolds, Monroe Batchelor, Tommy Murdock, Andrew Reynolds, and Valentine Batchelor. (Courtesy of Grant County Museum.)

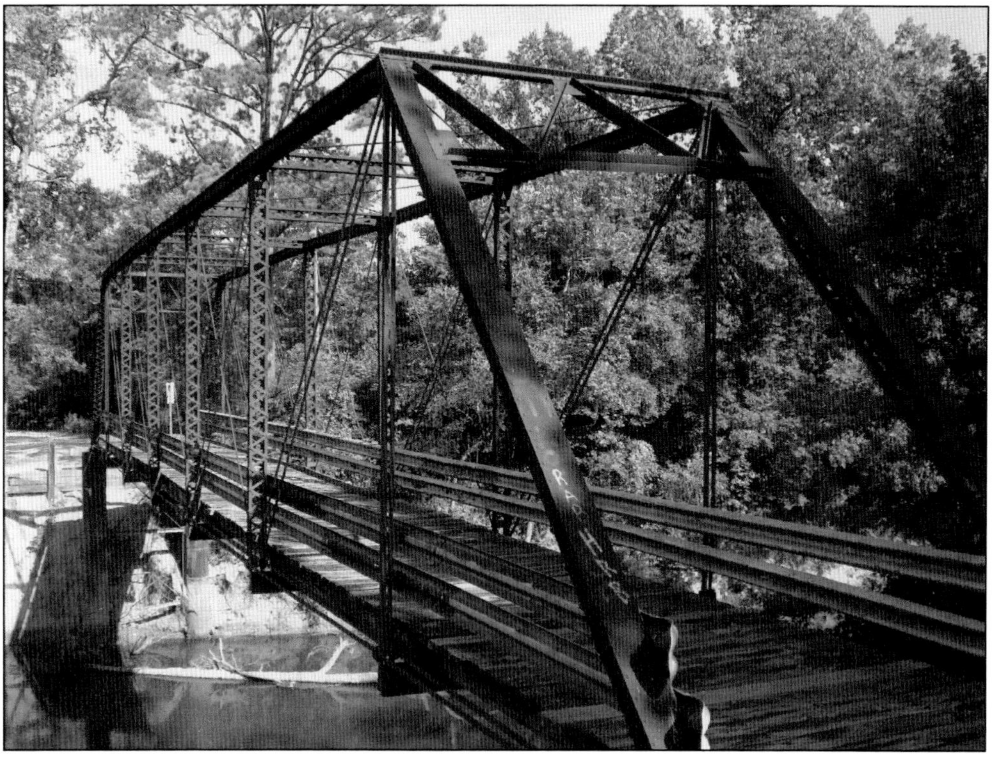

The last surviving iron bridge of Grant County provided service for a century, crossing the Saline River west of the community of Tull. Replaced in 2005 by a modern bridge, it stands silent witness to a much different era. (Courtesy of the author.)

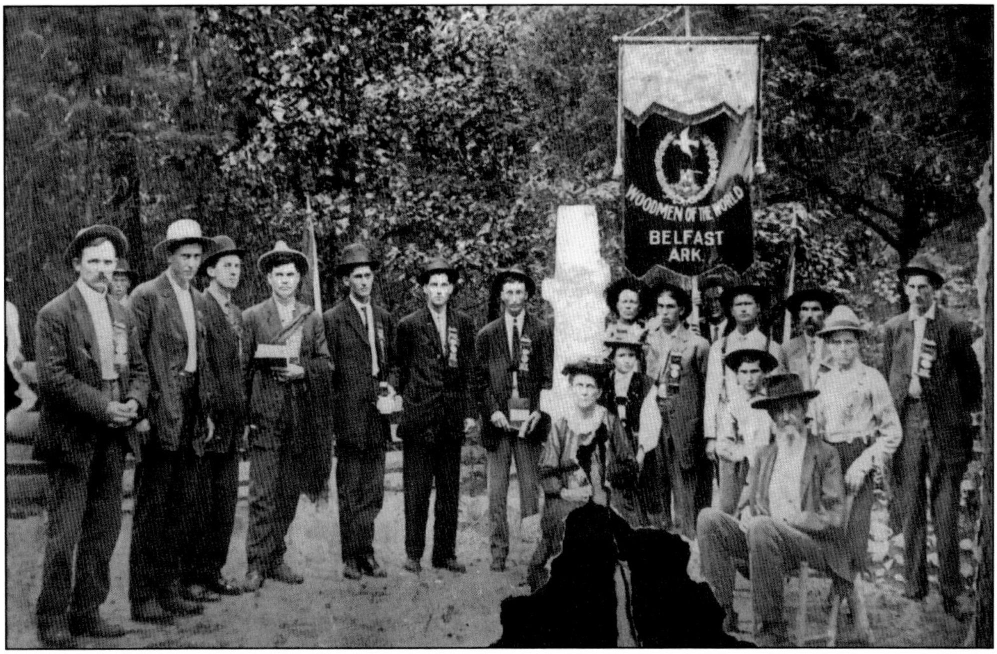


Woodmen of the World offered insurance in the early 1900s and provided cemetery markers for members such as the Martin family. At a dedication in Belfast in northwestern Grant County are, from left to right, (standing) two unidentified, Robert Pierce, George Hicks, Frank Wells, George Gillis, Horace Caple, Virgil Morton, unidentified, Lawrence Jones (holding banner), two unidentified, and Lonnie Smithers; (in front) Mandy Martin (seated), Emma Gober, Willie Gober, Russ Martin (seated), and Steve Martin. (Courtesy of Grant County Museum.)

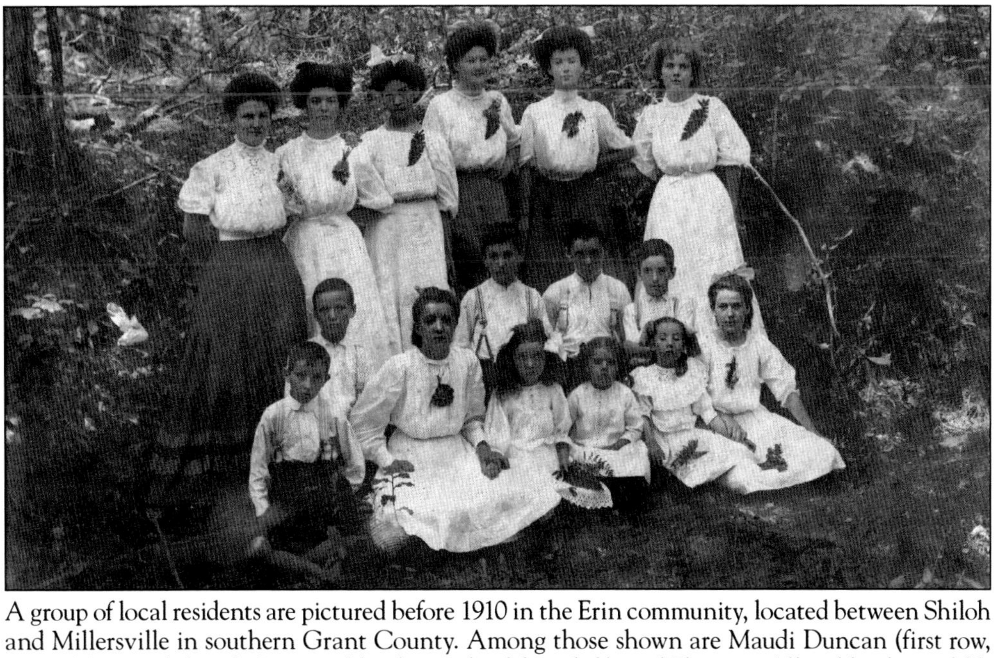

A group of local residents are pictured before 1910 in the Erin community, located between Shiloh and Millersville in southern Grant County. Among those shown are Maudi Duncan (first row, second from right), Lela Bratton (third row, third from left), and Florence Allen (third row, fifth from left). (Courtesy of Grant County Museum.)

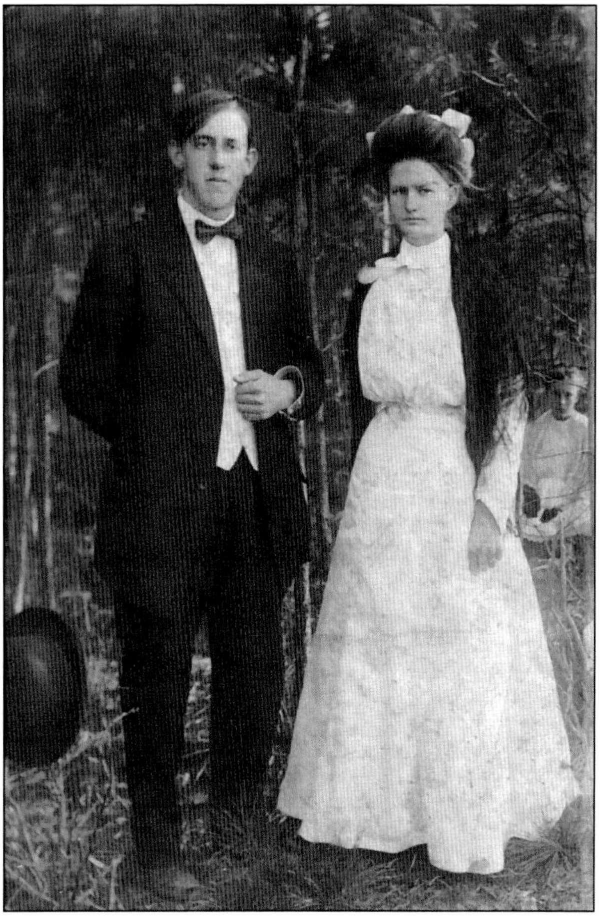

Will and Mattie Rhodes are pictured on their wedding day in December 1909. They would go on to raise five sons: William Cecil, James Clinton, Jethro Milton, Hugh Bentley, and Moffett Richard Rhodes. Will was a grandson of Dr. Richard Clinton and Susan Davis Rhodes. Mattie, eldest daughter of Hugh Randolph and Nancy Elizabeth Rogers Wilson, later divorced Will and lived out her life in Red Bluff, California. (Courtesy of the author.)

Posing with hats and cards are, from left to right, Effie Koon, Dessie Shell, Midge Wilson, Bessie Shell, Ed Wilson, and Bud Wilson. Millersville, where this photograph was taken, was a thriving community in 1910 due to the cutting of virgin bottomland timber. Hundreds of workers lived in small huts that were delivered by train. (Courtesy of the author.)

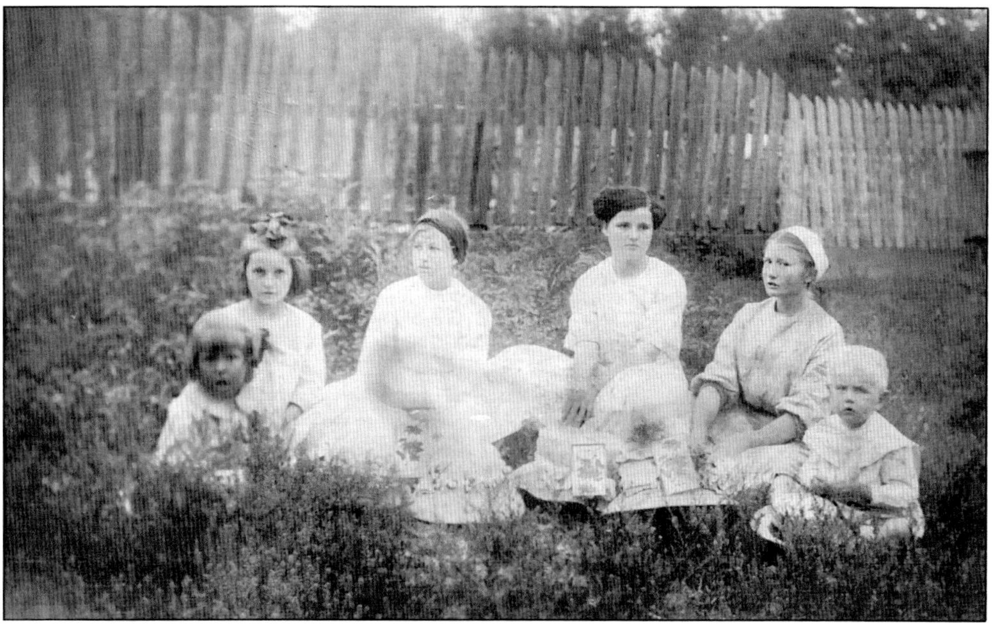

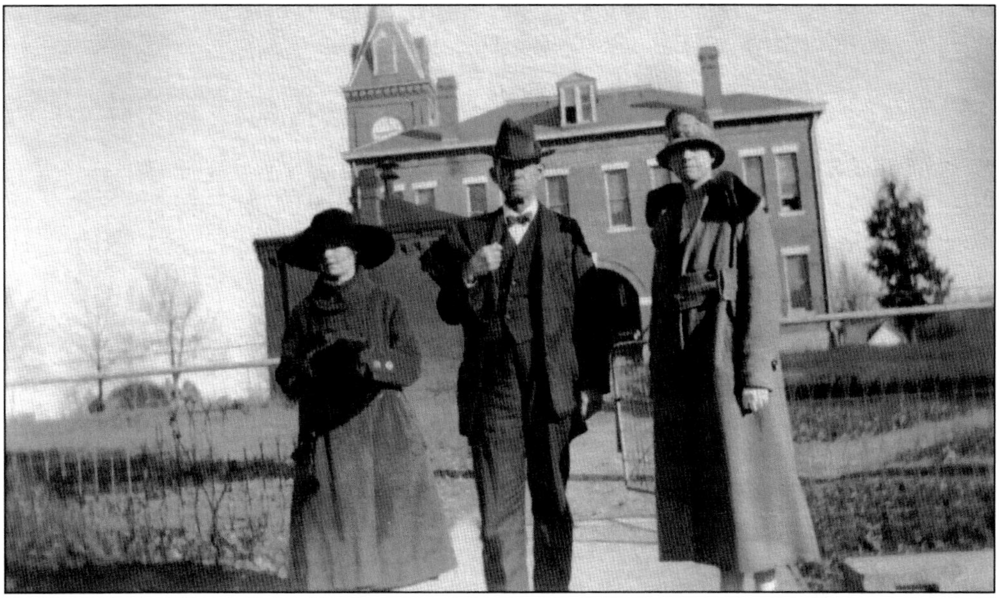

Among the items in the Dr. John W. Harrison collection is this photograph capturing the west side of the 1909 Grant County Courthouse and the first jail in Sheridan. (Courtesy of Grant County Museum.)

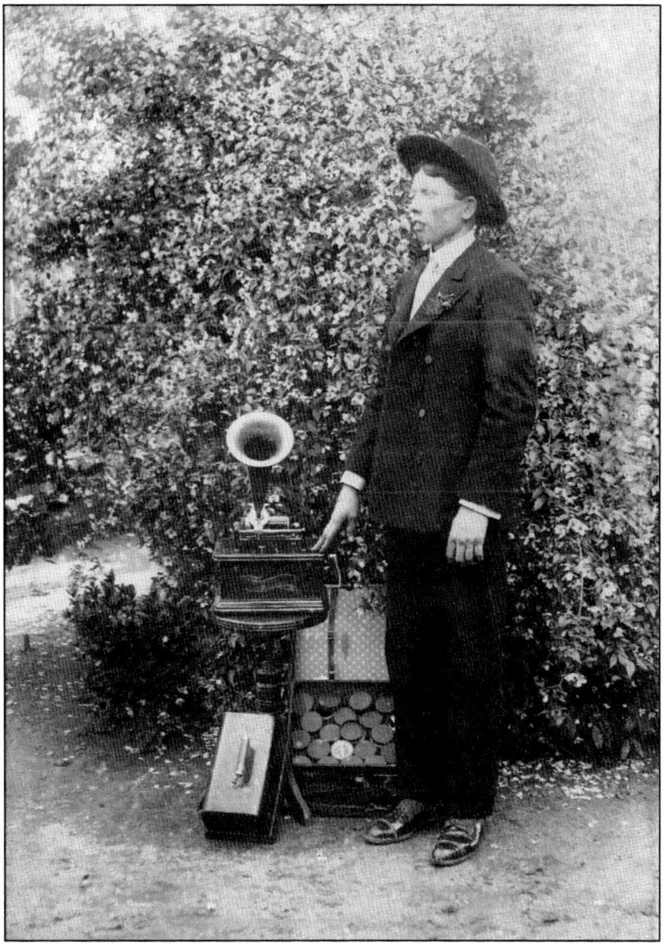

Virgil Walker is shown here selling gramophones in Grant County in the early 1900s. (Courtesy of Hazel Wilson.)

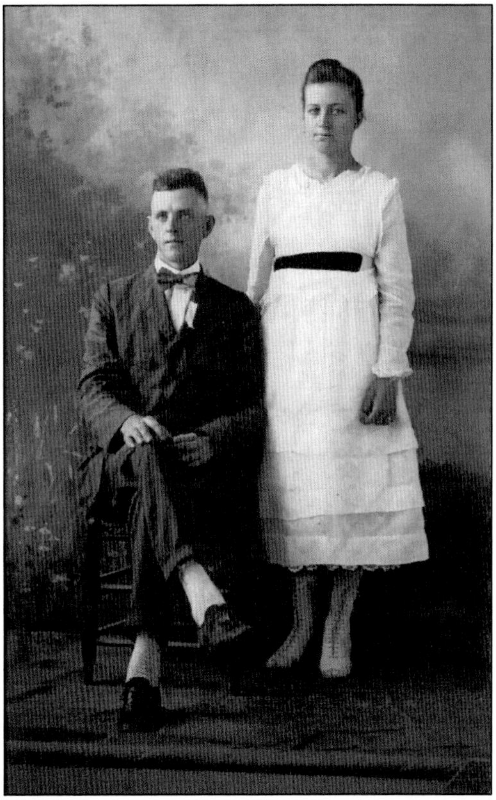

Girven and Lois Lybrand Brewer lived in southern Grant County. Girven, son of Pink and Lavada Bratton Brewer, served in France during World War I, then worked in the timber industry. Lois was the daughter of Will and Ida Hamilton Lybrand. (Courtesy of Hazel Wilson.)

It was hot work in the summer and cold work in the winter. Ed Gean operated this sawmill near Sheridan in the early 1900s. The mill was twice destroyed by fire. (Courtesy of Grant County Museum.)

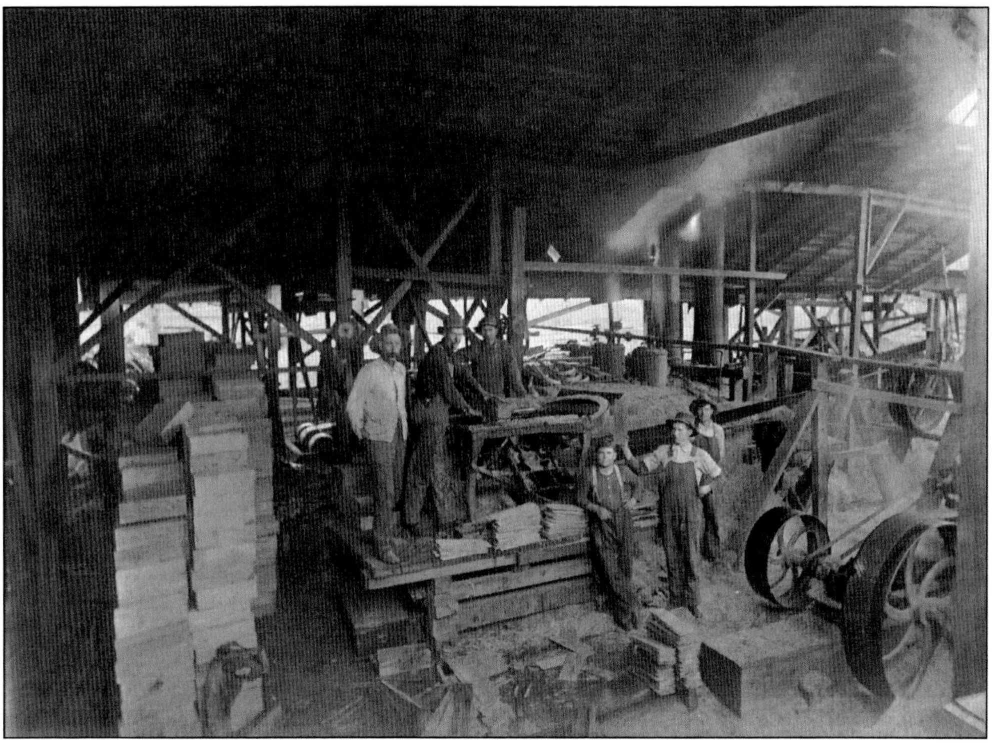

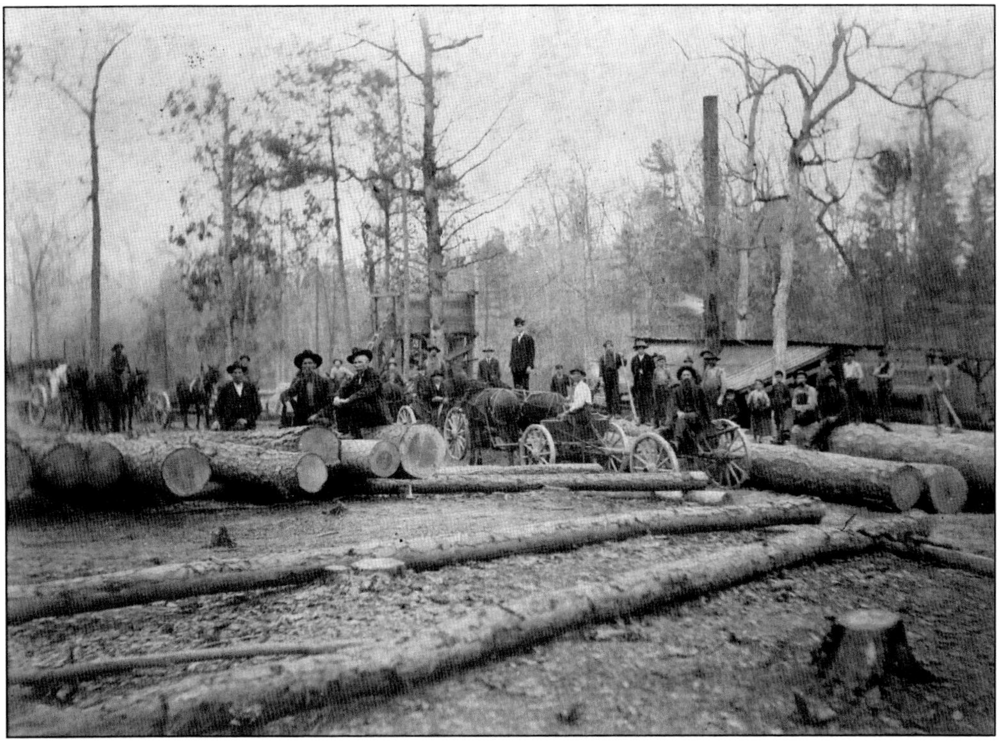

The Harrison Logging Camp employed several workmen near Leola to send timber to market in the early 1900s. (Courtesy of Grant County Museum.)

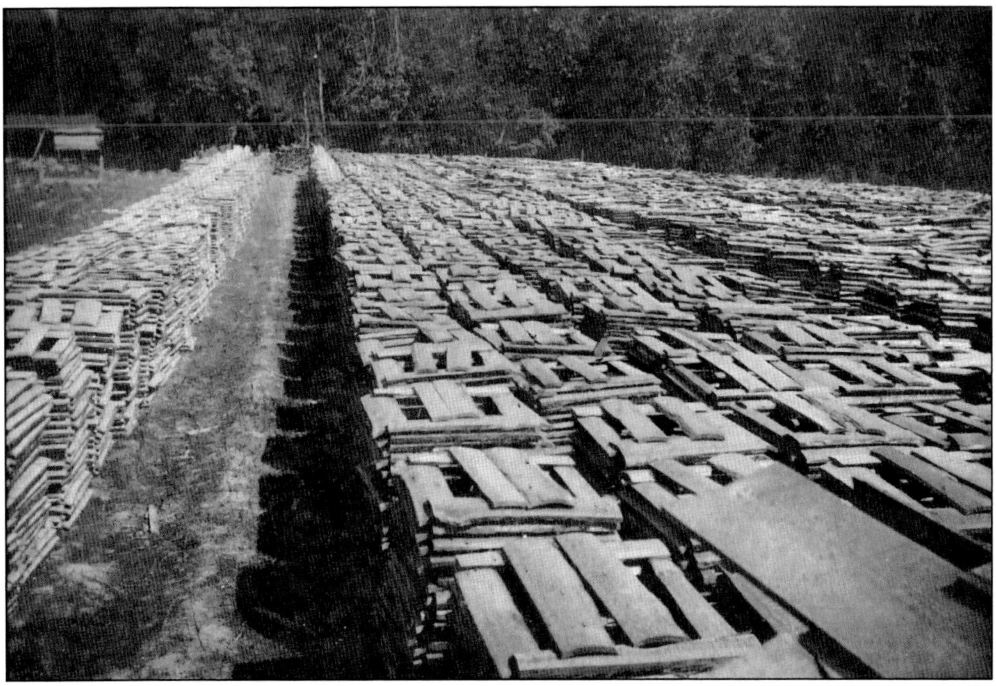

Providing staves for thousands of barrels is this carefully stacked stockpile at Leola in the early 1900s. (Courtesy of Grant County Museum.)

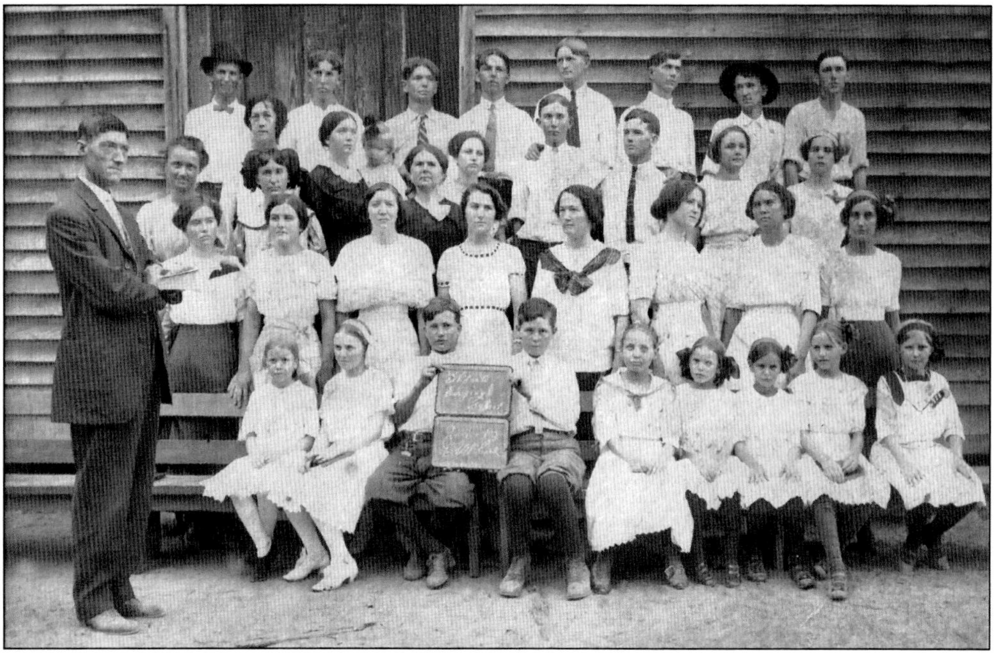

The Shiloh Singing School, with teacher Ed McCool, is pictured on September 4, 1913. Among those shown here are the following: (first row) Davis Wilson (third from left, holding slate) and Merle Henslee Reynolds (sixth from left); (second row) Della Henslee Wilson (third from left), Beatrice Barnes (sixth from left), and Alice "Big Sis" Henslee Camp (seventh from left); (third row) Sally Allen Ward (at left) next to her sister, Minnie "Cooter" Allen Burgess, Lela Bratton Moore (behind Burgess), Florence (holding child), and Lawrence Wilson (fourth from right, with a hand on his right shoulder); (fourth row) "Little" Tom Allen (fourth from left), Lonnie Gartman (sixth from left), Rich Allen (seventh), and Carl Wilson (eighth). (Courtesy of Hazel Wilson.)

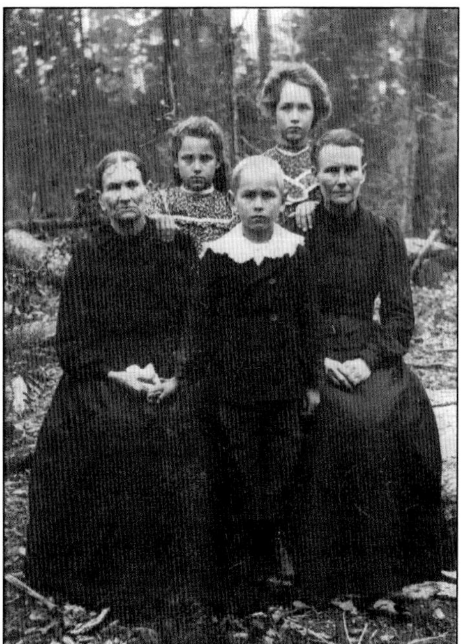

Eliza Black Bratton Hood (left) arrived in Arkansas in the 1840s with her parents, Joseph S. and Jane Handy Black, natives of South Carolina. She poses here with her daughter Matilda Jane Bratton Barnes (far right) and her grandchildren, from left to right, Lou (born 1892), Berton (born 1895), and Iris Barnes (born 1890). Their mother, Mary Ella Warlick Barnes, died in 1899. (Courtesy of the author.)

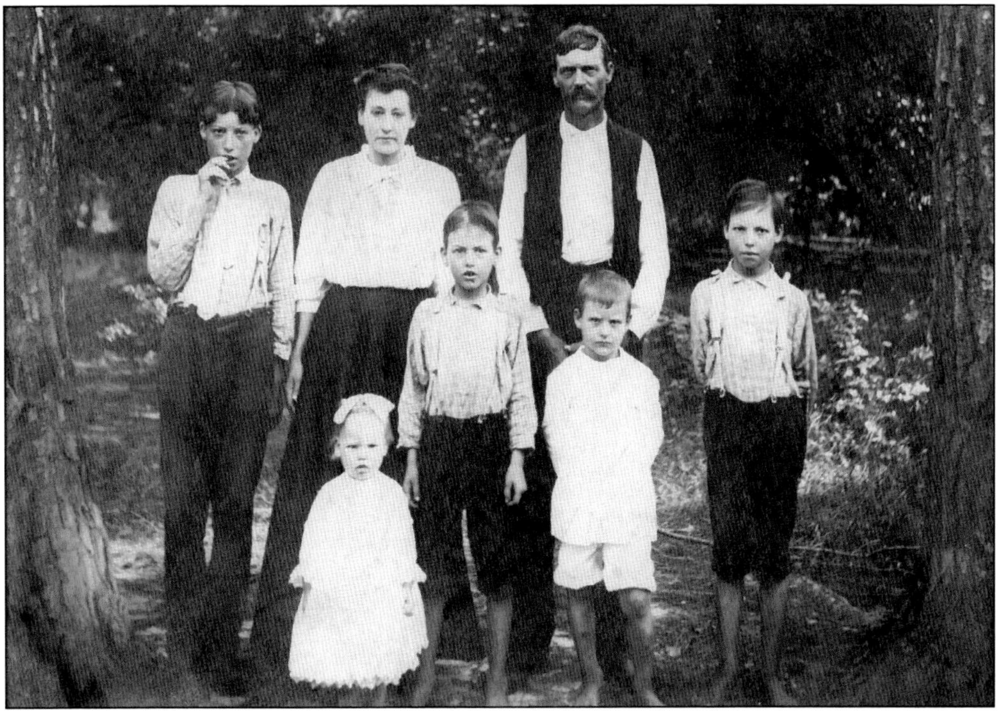

The Samuel Daniel Byrd Jr. family, seen here in 1913, lived near the Hot Spring and Grant County line, west of Poyen. The family members are, from left to right, (first row) Madge Cleo Byrd, David Hugh Byrd, Martin D. Byrd, and John Wesley Byrd; (second row) William Hobart Byrd Sr., Susan Amanda Ballard Byrd, and Samuel Daniel Byrd Jr. Madge graduated from Sheridan High School in 1931 and remained at the old home place all of her life. (Courtesy of W.H. Byrd Jr. family.)

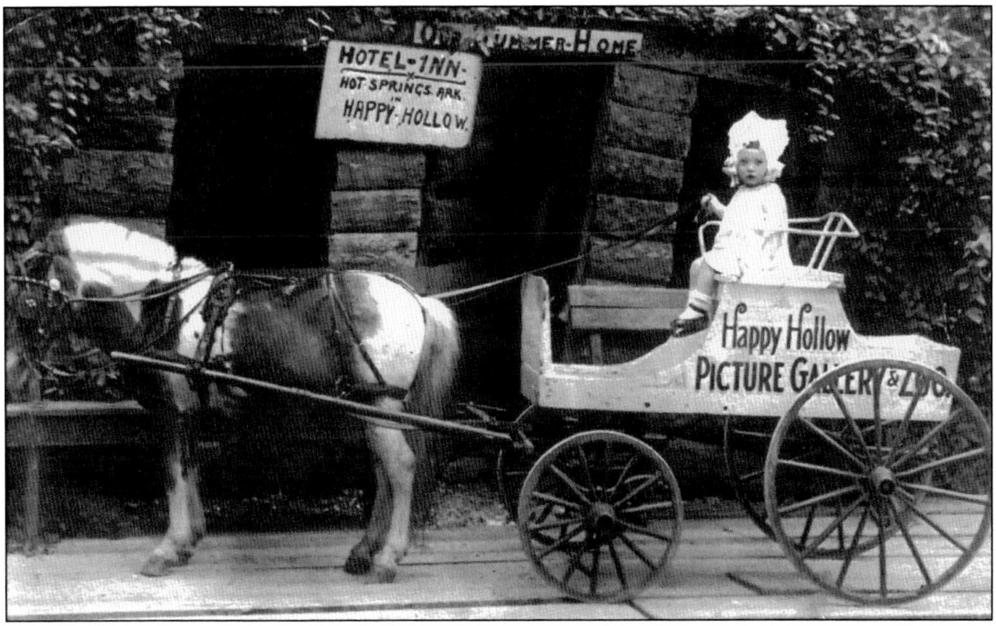


Sheridan's Dr. O.W. Hope and popular English teacher Alice Bland Hope's daughter Louise is shown here in an Easter bonnet at Hot Springs. (Courtesy of Grant County Museum.)

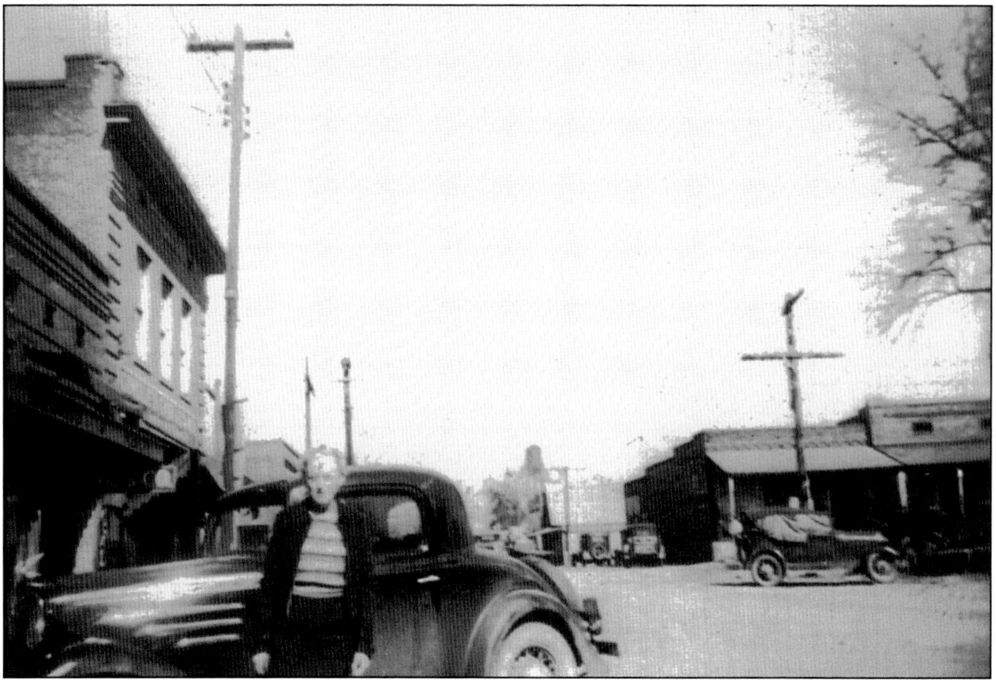

Clyde Walker arrives on Sheridan's Oak Street, perhaps for shopping, in the early 1930s. (Courtesy of Grant County Museum.)

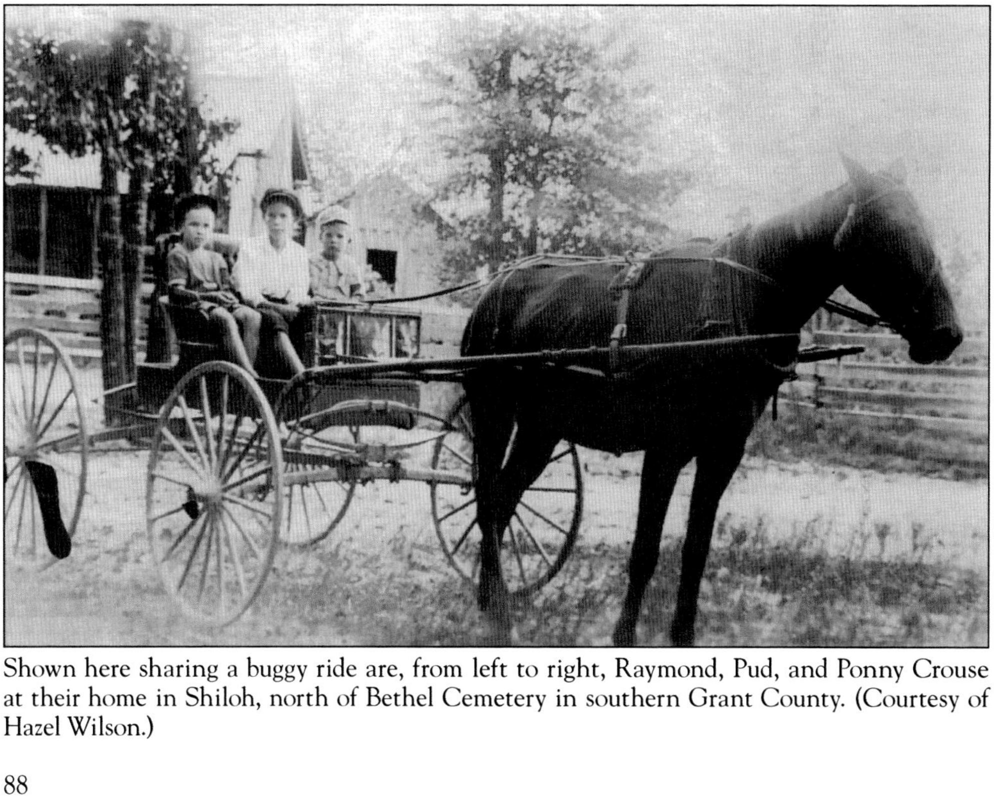

Shown here sharing a buggy ride are, from left to right, Raymond, Pud, and Ponny Crouse at their home in Shiloh, north of Bethel Cemetery in southern Grant County. (Courtesy of Hazel Wilson.)

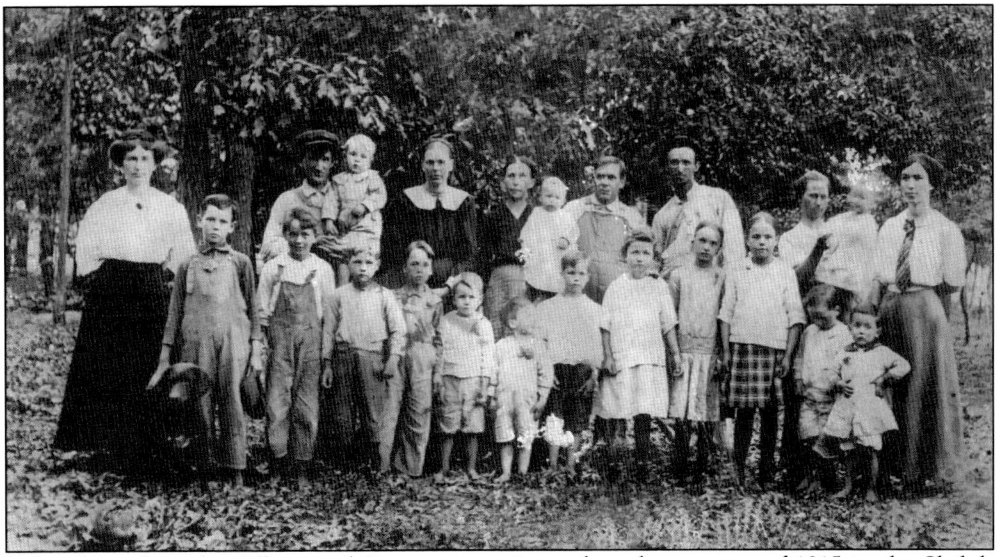

Related families, the Allens and Crouses, are pictured in the summer of 1915 in the Shiloh community of southern Grant County. Shown here are, from left to right, the following: (first row) Clyde Crouse, Ivan Allen, Jess Hunter, Doyle Allen, Jewel Allen, Raymond Crouse, Pud Crouse, Myrl Allen, Wilma Allen, Doris Crouse, Buck Allen, and Olen Acree; (second row) Minnie "Cooter" Allen Burgess, Rich Allen (holding son Boyce Allen), Lizzie Bratton Allen, Lizzie Bratton Crouse (holding Ozra "Ponny" Crouse), Johnny Crouse, "Shorty" Allen, Rose Walker Allen (holding son Tootie), and Florence Allen Acree. (Courtesy of the author.)

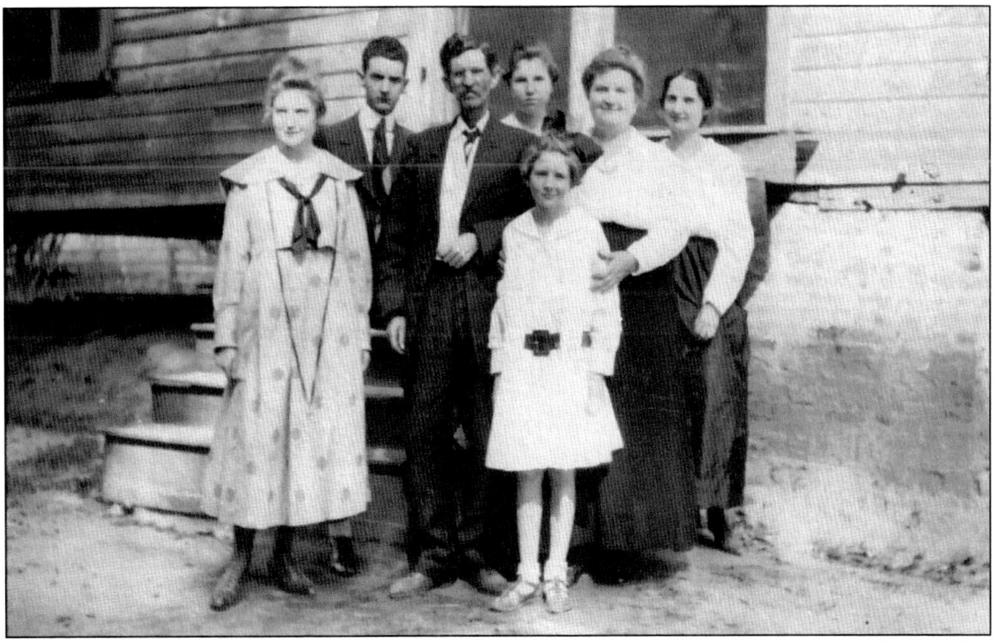

The Fullers of Princeton, posing here for a photograph, are, from left to right, Ann Bolling Fuller, Al Fuller, Will Fuller, Agnes Fuller, Flournoy Fuller (in white dress), Virginia Elizabeth "Jennie" Harrison Fuller, and Virginia Elizabeth Fuller Holmes. During the Civil War, Will hid his horse as Union troops approached, later explaining that "he was too young to fight, but old enough to get shot." (Courtesy of Elizabeth Ann Lancaster and Grant County Museum.)

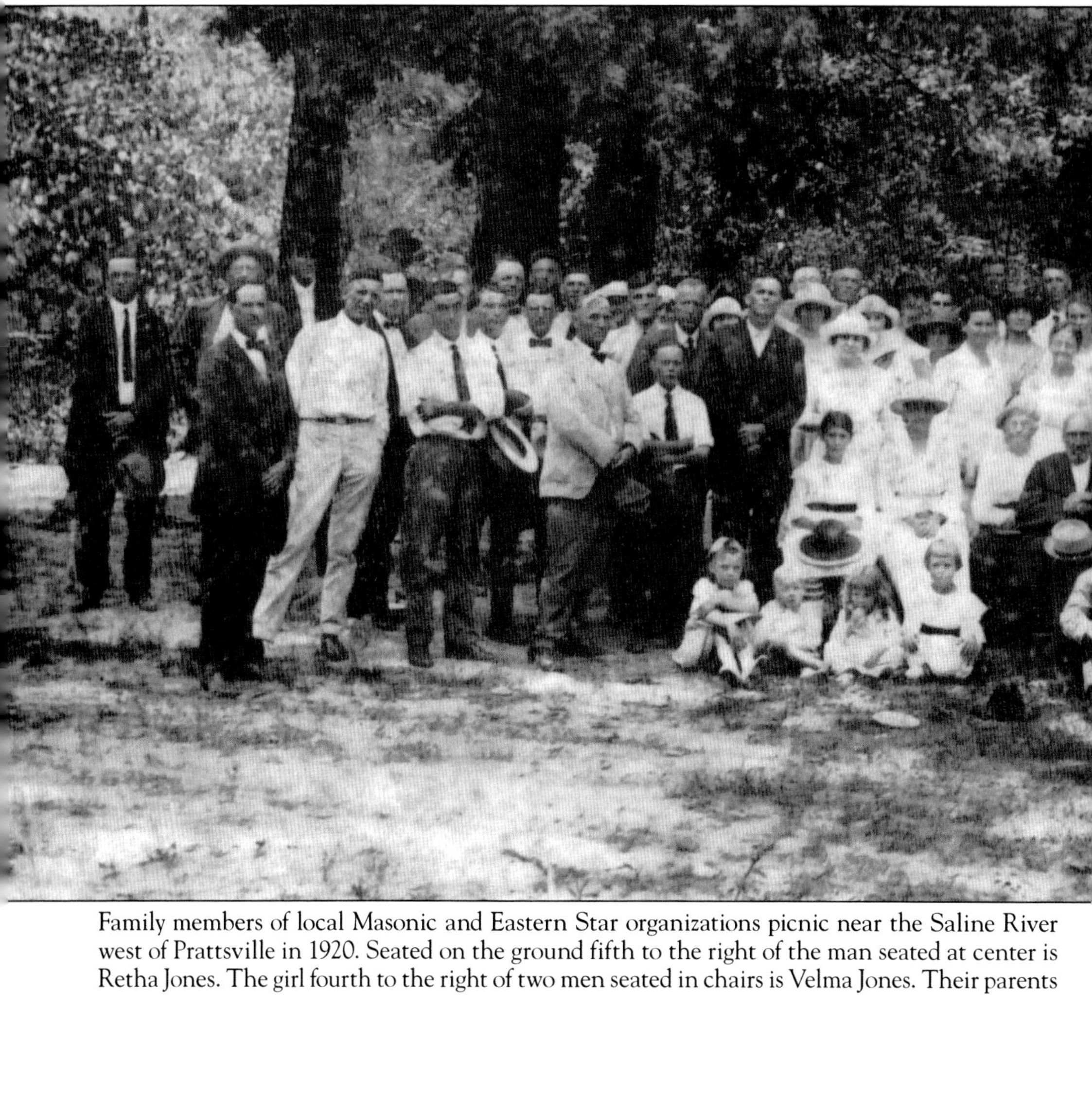

Family members of local Masonic and Eastern Star organizations picnic near the Saline River west of Prattsville in 1920. Seated on the ground fifth to the right of the man seated at center is Retha Jones. The girl fourth to the right of two men seated in chairs is Velma Jones. Their parents

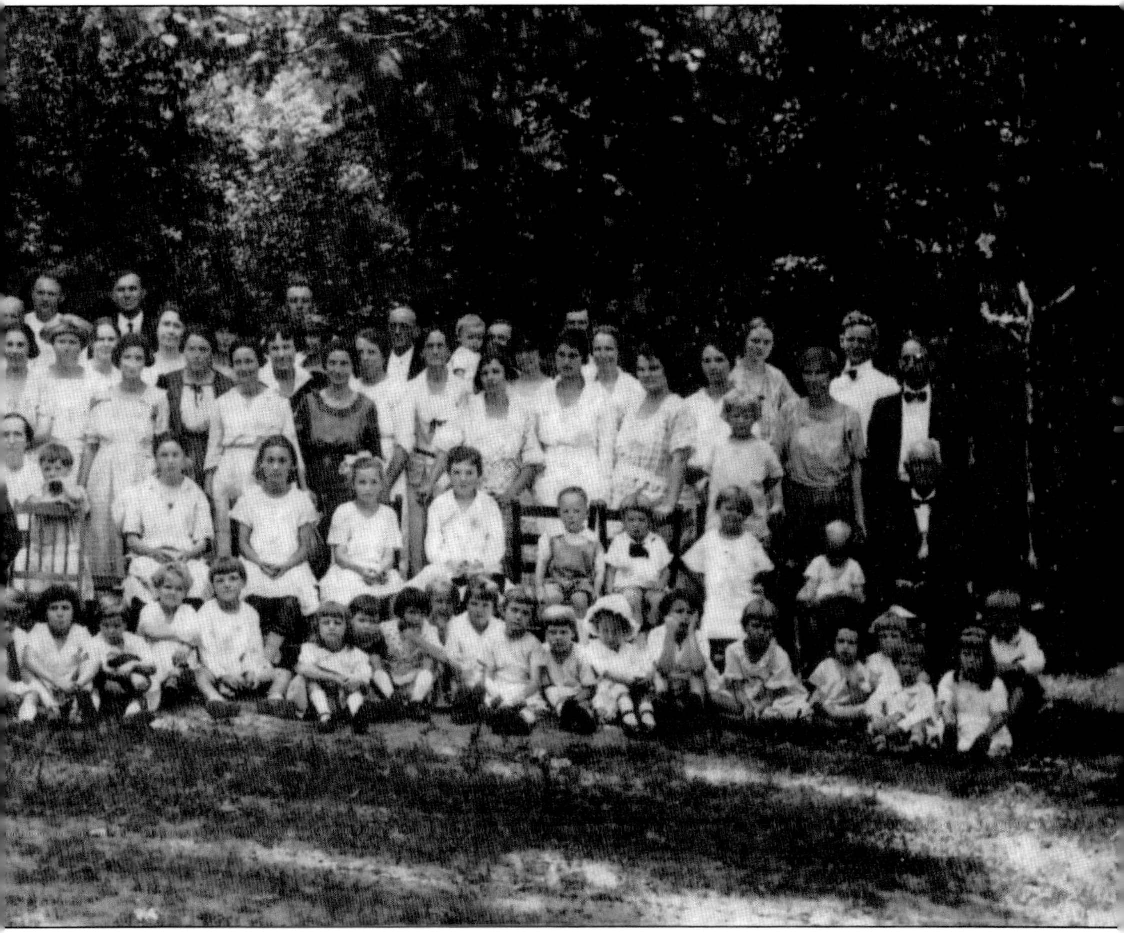

are Claudia Reese Jones (standing left of Velma in white dress, partially hidden behind the left shoulder of a lady in a dark dress) and Lem Jones (standing to the right of the empty chair in the back row with a dark tie). (Courtesy of Doug Lloyd.)

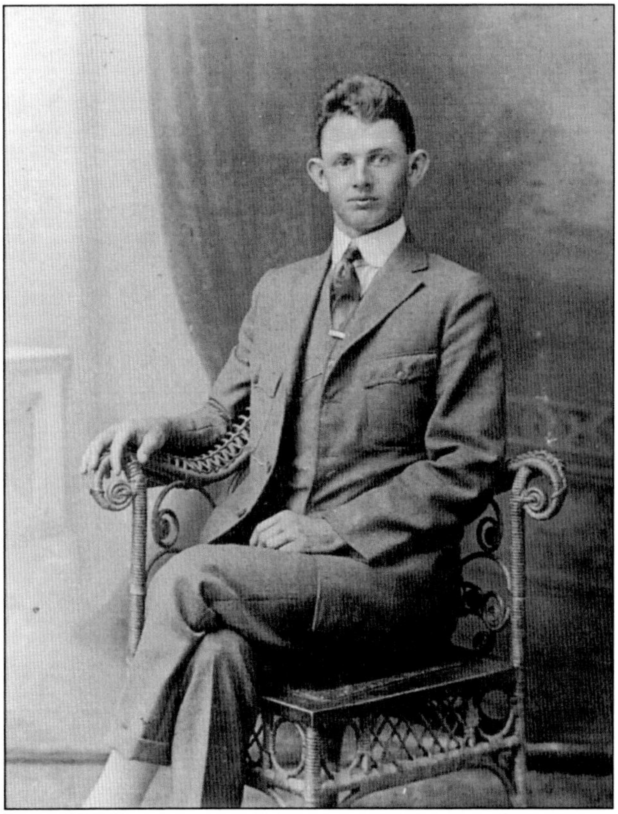

Grant County native Earl Brewer, son of Pink and Lavada Bratton Brewer, is pictured here in the early 1900s. Earl's brothers were Girven, Otis, and Scott, and his sisters were Altus, Lillian, Ruth, and Elsie. His wife was Mildred Lites Brewer, and they lived in the Pine Bluff area. (Courtesy of Hazel Wilson.)

Standing in the T.C. Rushing store in Sheridan are, from left to right, Earl Rushing, Mary Cabiness, Zella Rushing, and Hosea Shepherd. (Courtesy of Grant County Museum.)

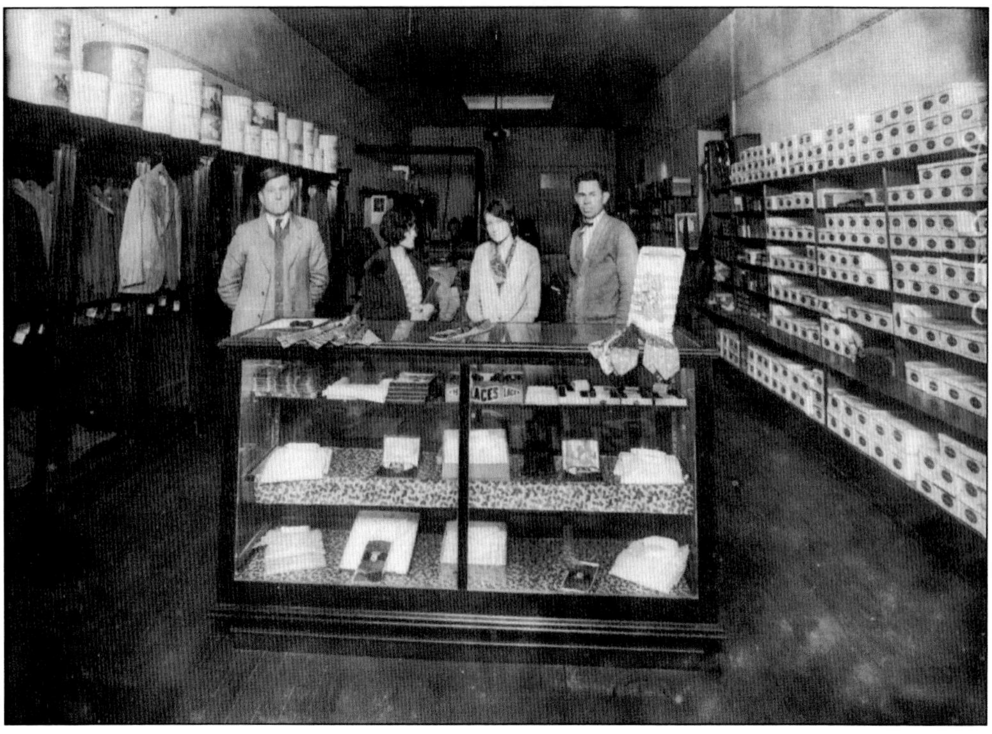

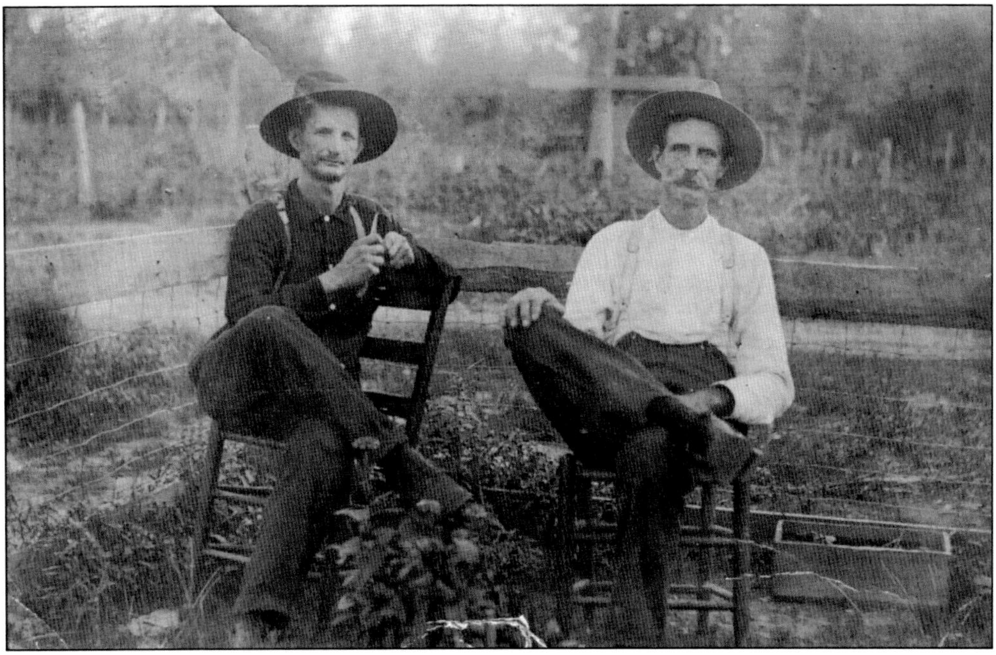

Horace Montgomery Wilson (left) and his brother Hugh Randolph Wilson were born south of the Saline River to Andrew and Martha Jane Wilkinson Wilson. They were named after their grandfather, Hugh Montgomery Wilkinson, a native of South Carolina. "Uncle Dolph" made fine sorghum molasses. (Courtesy of the author.)

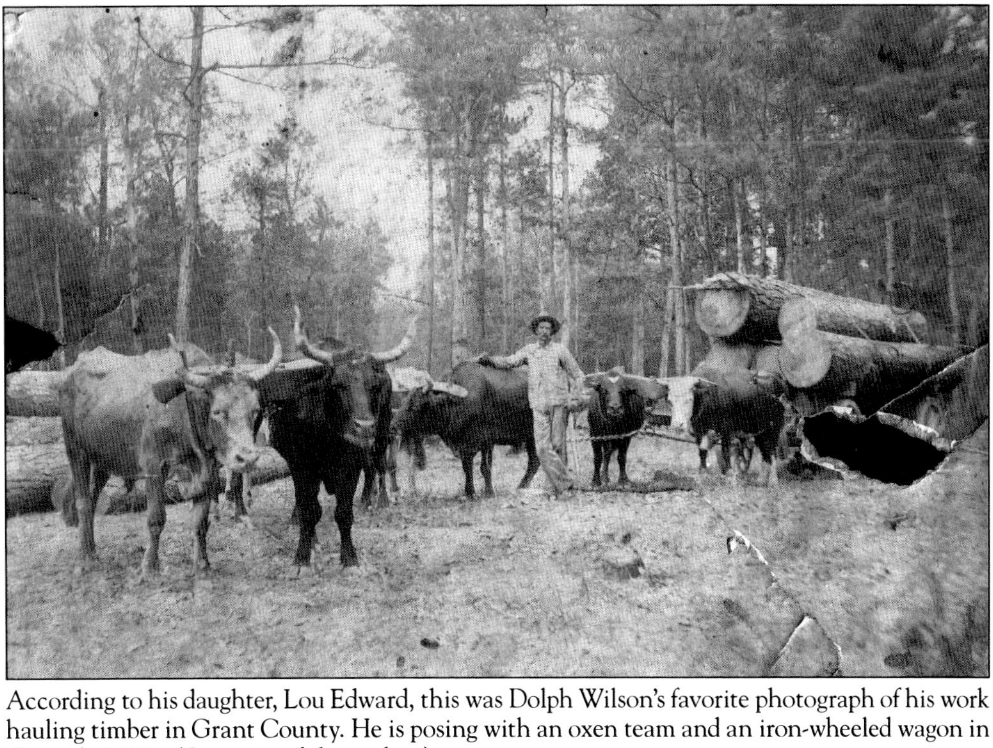

According to his daughter, Lou Edward, this was Dolph Wilson's favorite photograph of his work hauling timber in Grant County. He is posing with an oxen team and an iron-wheeled wagon in the early 1900s. (Courtesy of the author.)

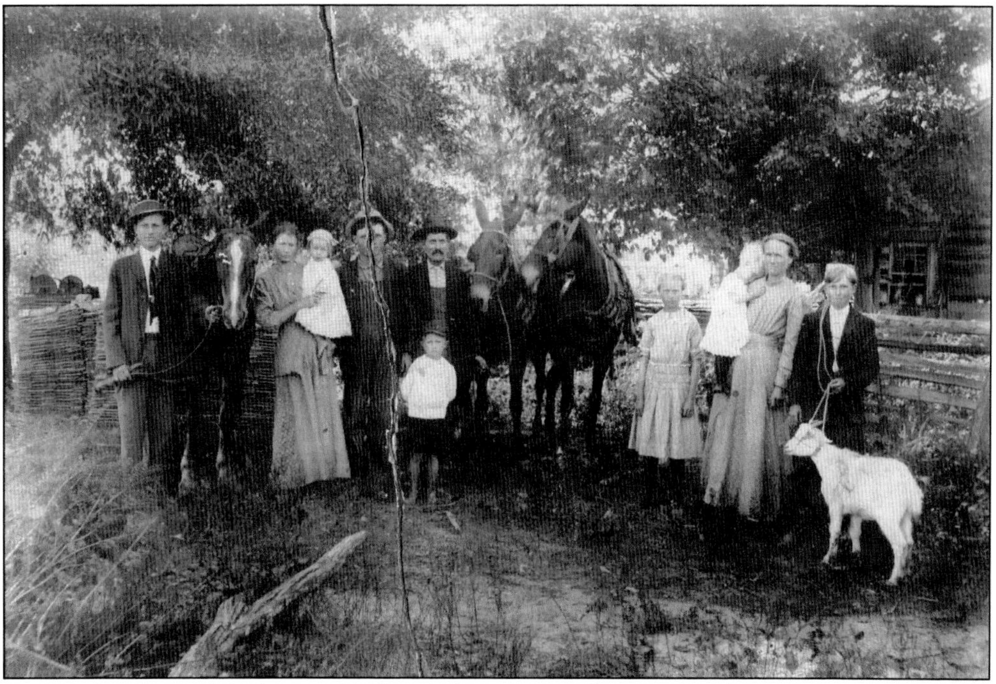

Gathered in front of the home of Jeff and Ida Barnes Wilson in the Shiloh community in 1915 are, from left to right, Davis Wilson, Lou Edward Wilson (holding son Havis Wilson), James Roy "Dink" Wilson (Lou's husband), Carl Wilson, Calvin Jefferson Wilson, Neville Wilson, Ida Barnes Wilson (holding daughter Alma), and Ralph Wilson. Carl, Neville, and Alma would live out their lives in and near Red Bluff, California. (Courtesy of the author.)

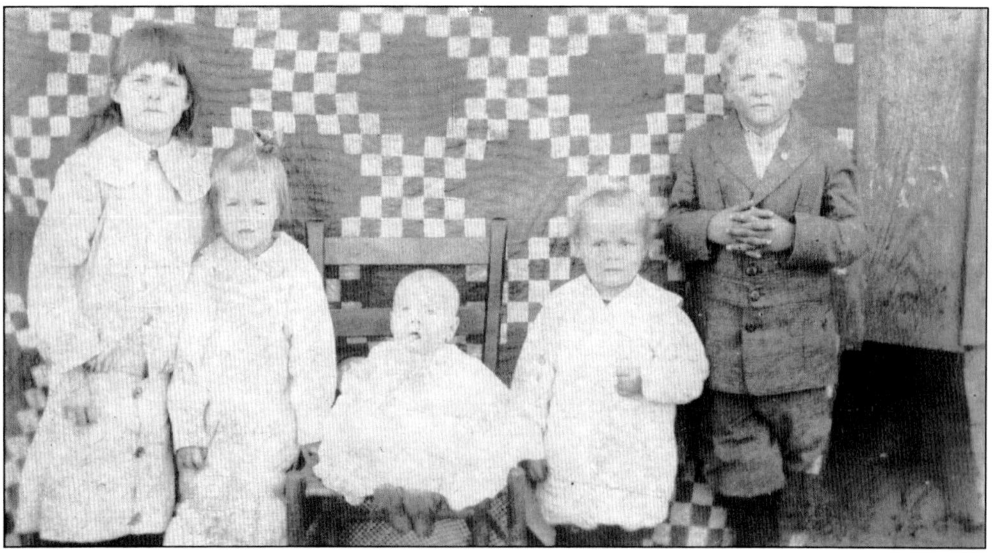

Shown here are, from left to right, Effie Koon and children of Hugh Randolph Wilson, Willie, Leo, Joe T., and Bud, pictured at their father's home in southern Grant County. (Courtesy of the author.)

Ora Byrnes and her husband pose in this early-1900s picture. A niece of Louisa Payne Bland, she was raised by her grandmother, Nancye Payne, a relative of Alice Hope. (Courtesy of Grant County Museum.)

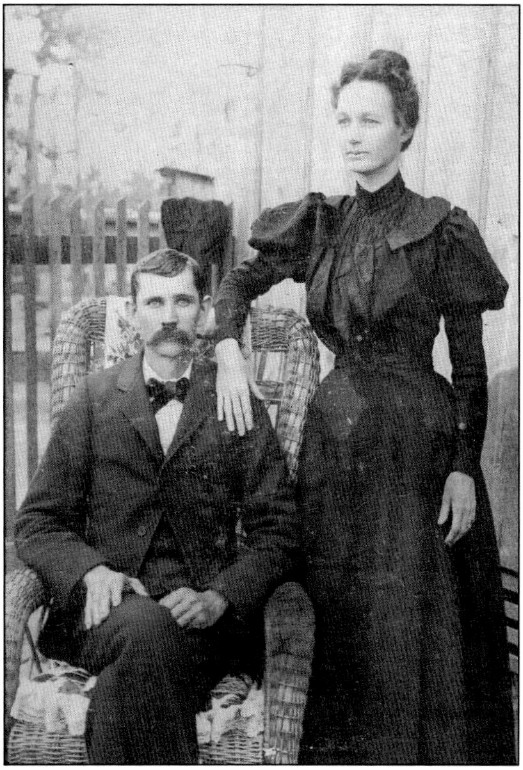

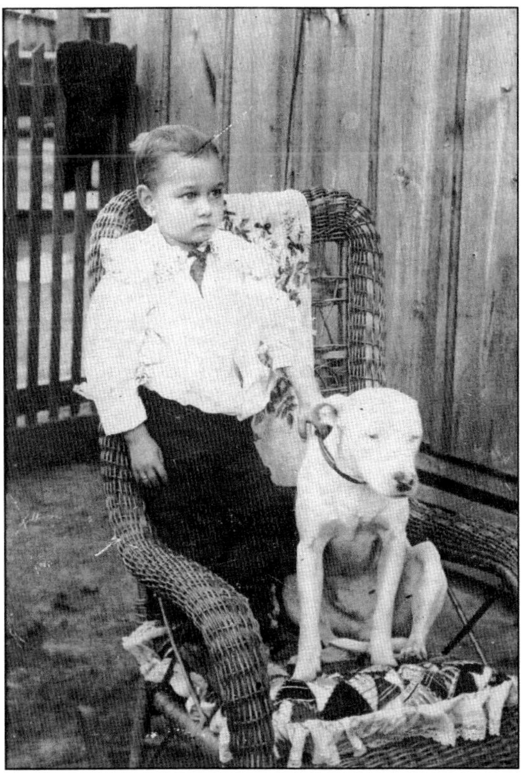

Ernest, son of Ora Byrnes, poses with his dog in the early 1900s in an Alice Hope photograph. (Courtesy of Grant County Museum.)

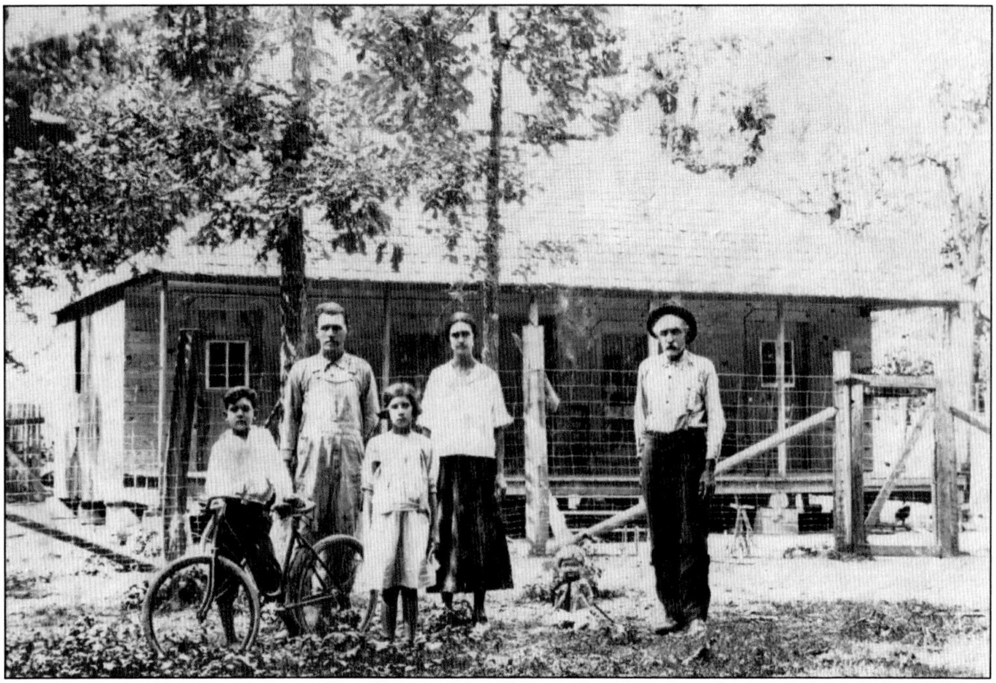

One mile south of Shiloh Church was the home place that Sam Allen settled in the 1870s with wife Josephine Berry Allen. Posing on the site are, from left to right, Olen Acree, Fred Allen, Clova Nell Allen (his daughter), Fannie Martindale Allen (his wife), and Samuel Wright Allen (his father). Samuel was born in 1852 in Carroll County, Tennessee. (Courtesy of the author.)

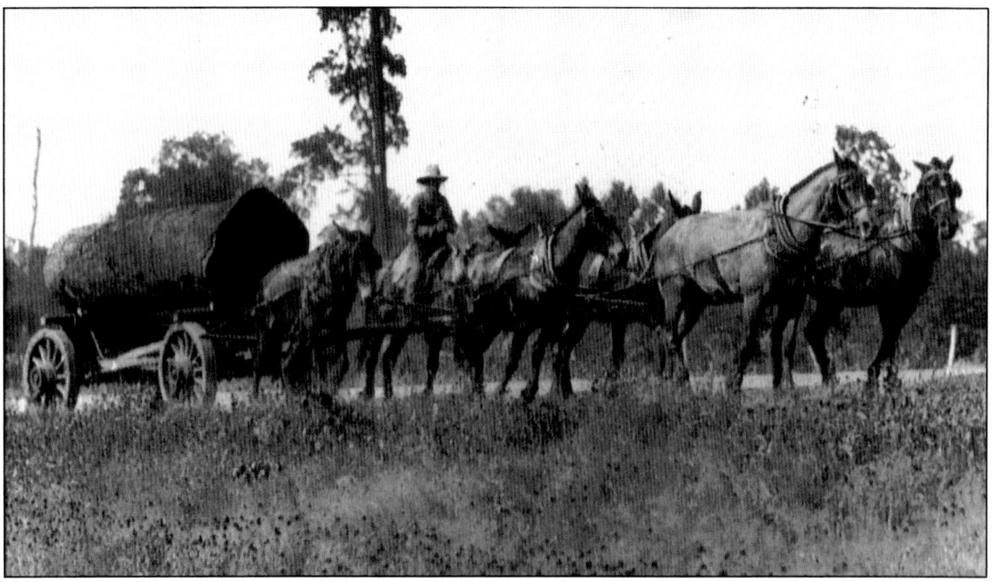

A local resident guides a team hauling a large oak log to a neighborhood sawmill. For most local people during the first half of the 1900s, work besides farming was either cutting, hauling, or sawing timber, which provided lumber for growing markets like Little Rock and Memphis. (Courtesy of Grant County Museum.)

Pearl House, first grandchild of Louise Payne Bland and J.D. Bland and "idol of the family," appears in a portrait once belonging to Alice Hope. (Courtesy of Grant County Museum.)

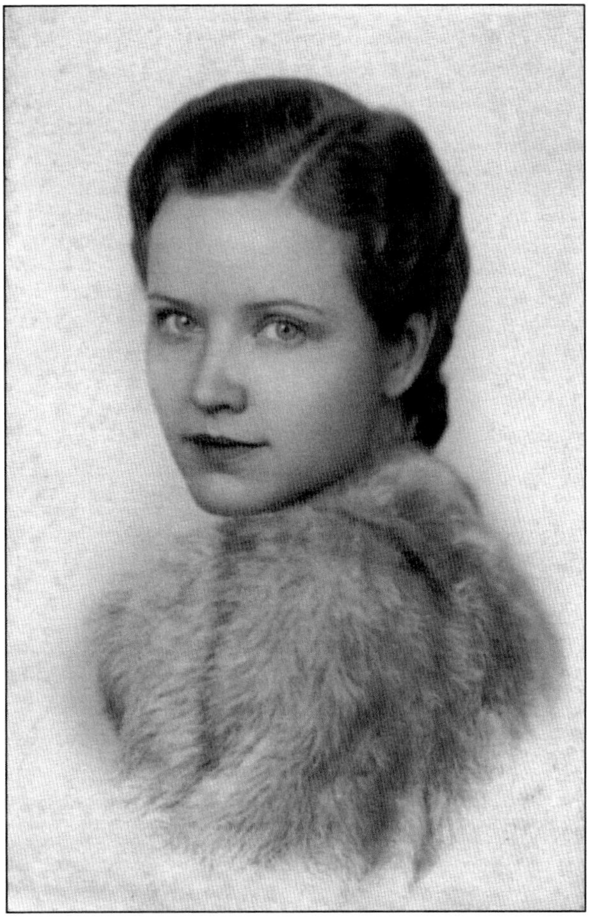

When built just prior to 1900 by the Henry Bradley and Clint Rushing families south of Grant County Bank, the Bradley-Rushing Building became a center of business at the southwest corner of Oak and Center Streets in Sheridan. Dental and law offices, mercantiles, restaurants, grocery stores, and the post office once occupied parts of the building. (Courtesy of the author.)

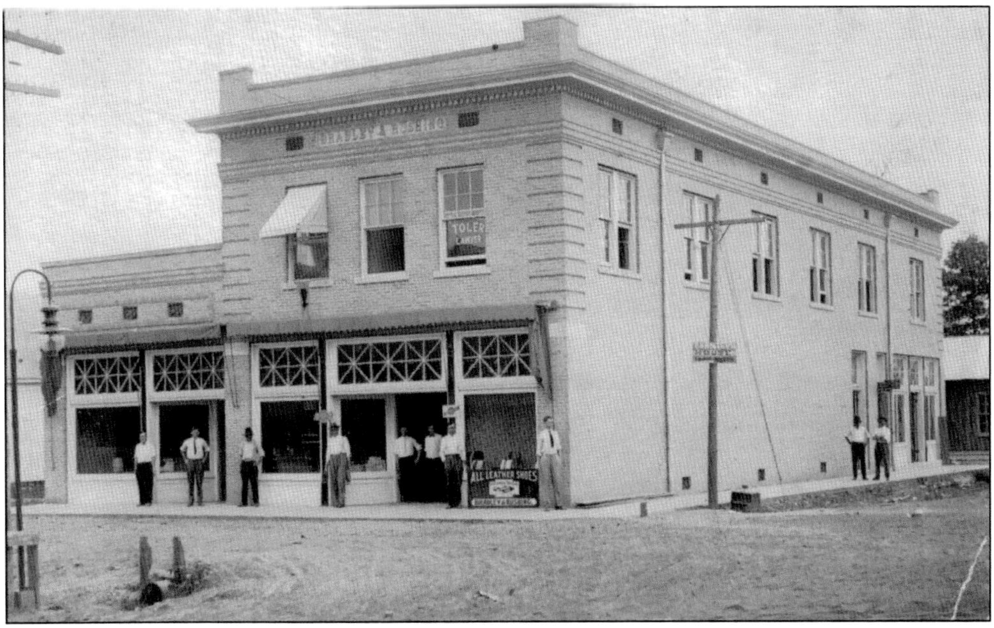

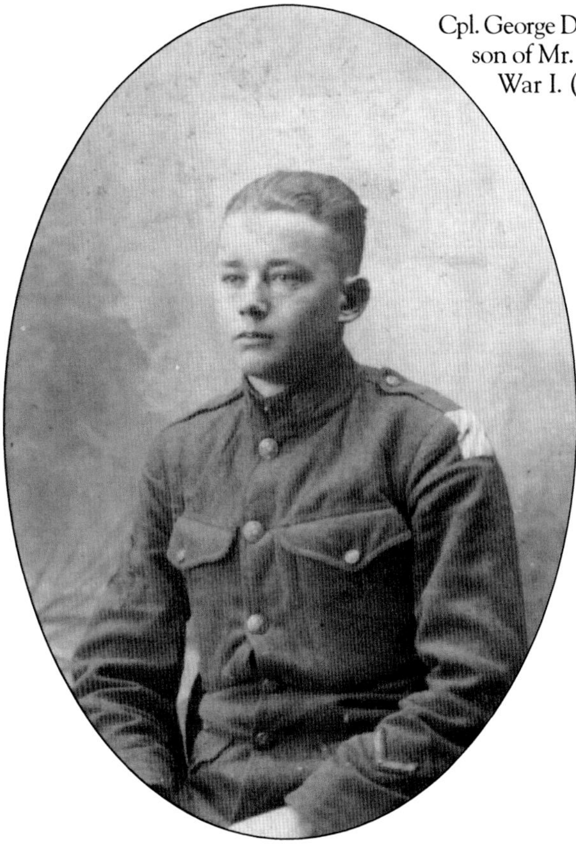

Cpl. George Dewey Bland, shown in this portrait, was the son of Mr. and Mrs. J.D. Bland. He served in World War I. (Courtesy of Grant County Museum.)

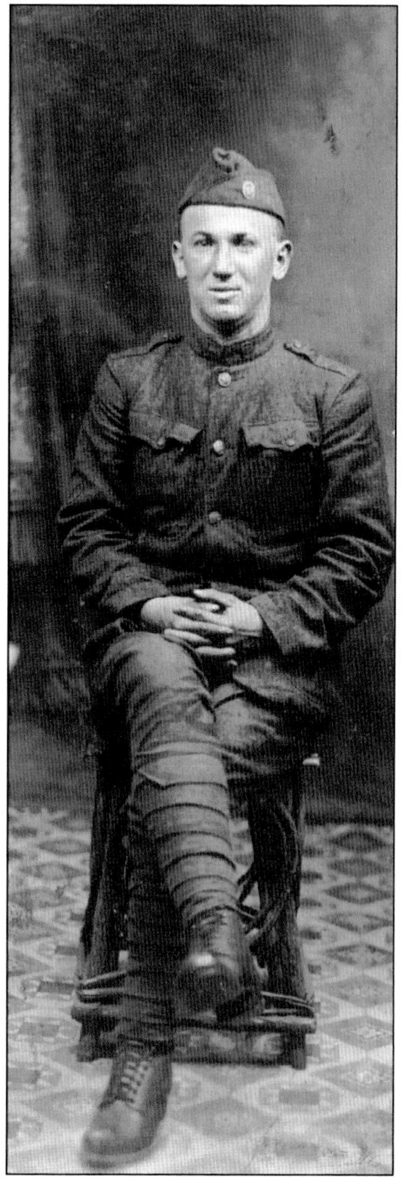

A local soldier, Willie, poses in his World War I uniform in this photograph, which is in the collection once owned by Retha Jones Johnson of Sheridan. (Courtesy of Doug Lloyd.)

Hobart Byrd (right) served as a medic in the 4th Division in France and the occupied Rhineland during World War I. He attended college, married Bertha Sheppard, and taught at schools in Poyen, Fenter, and Sheridan. Alongside him are unidentified soldiers from Wisconsin and Georgia. (Courtesy of W.H. Byrd Jr. family.)

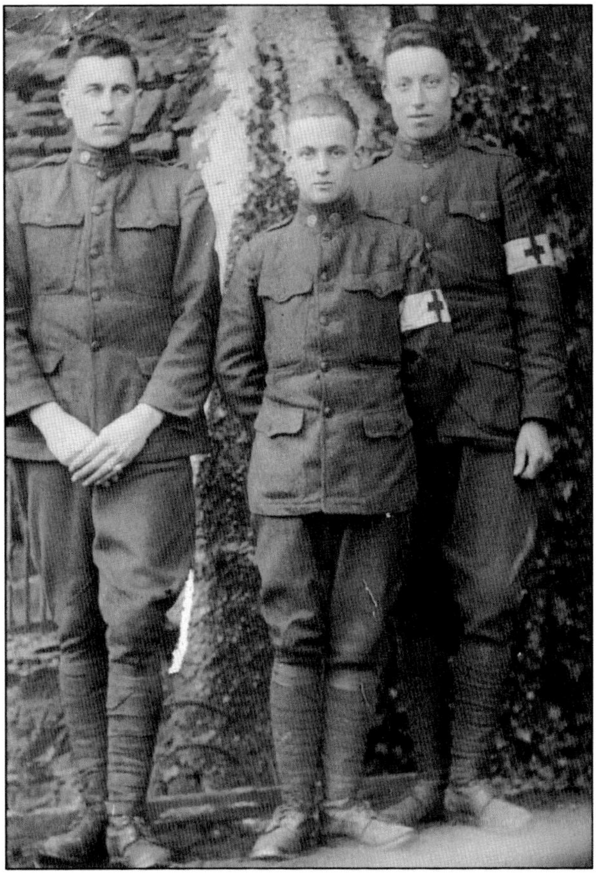

Addressed to Mr. R.A. Allen of Erin in Grant County and postmarked 6:00 p.m., July 27, 1918, from the Pike Branch, this postcard was sent by a local soldier reporting his arrival in the 64th Casual Company, 162nd Depot Brigade, Camp Pike, Arkansas. (Courtesy of Hazel Wilson.)

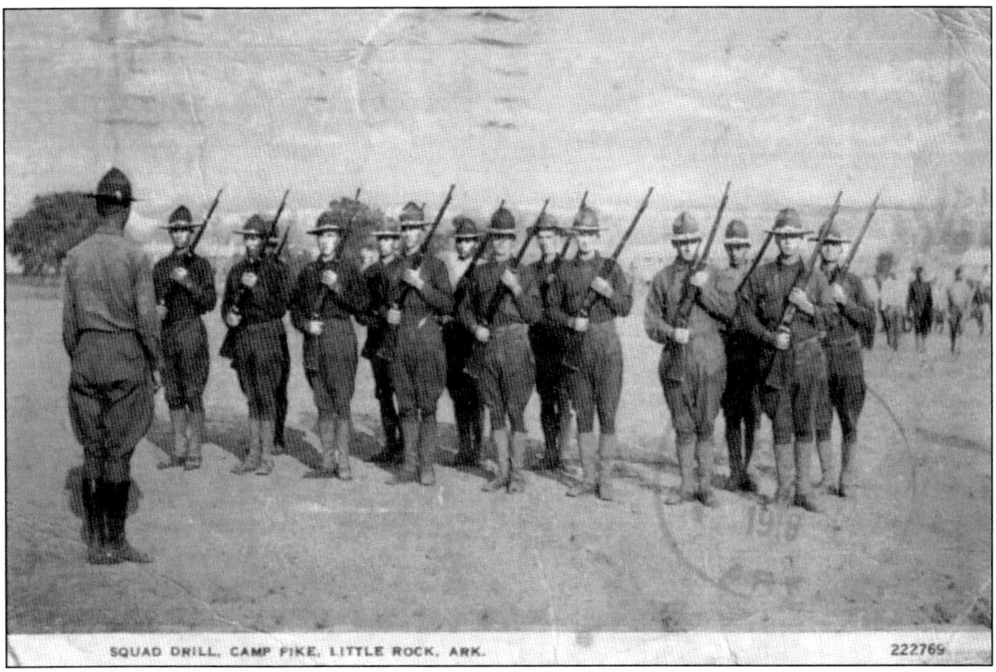

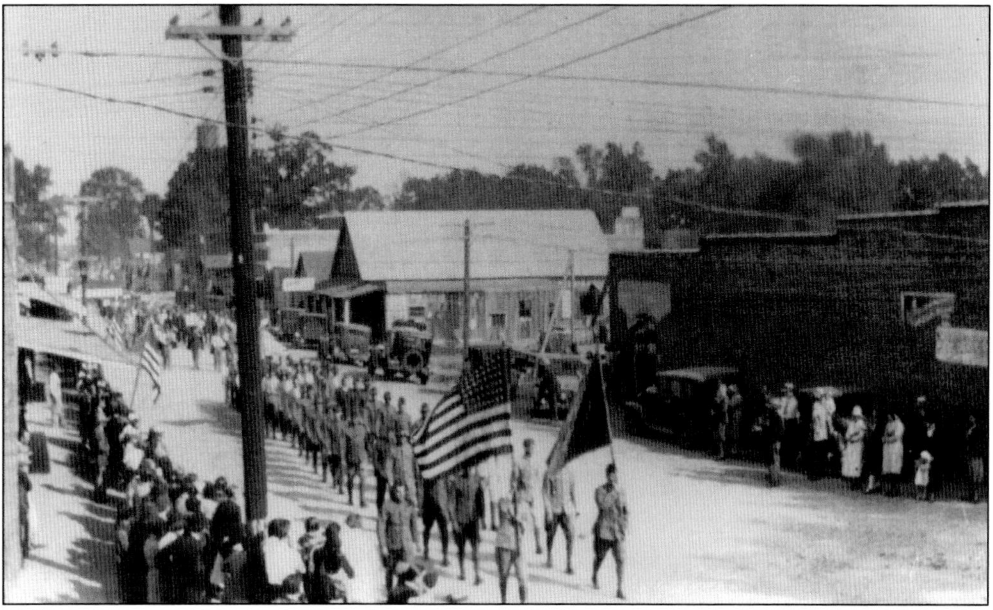

Soldiers from World War I lead an Armistice Parade down Oak Street before turning east onto Center Street north of the courthouse in Sheridan. (Courtesy of Grant County Museum.)

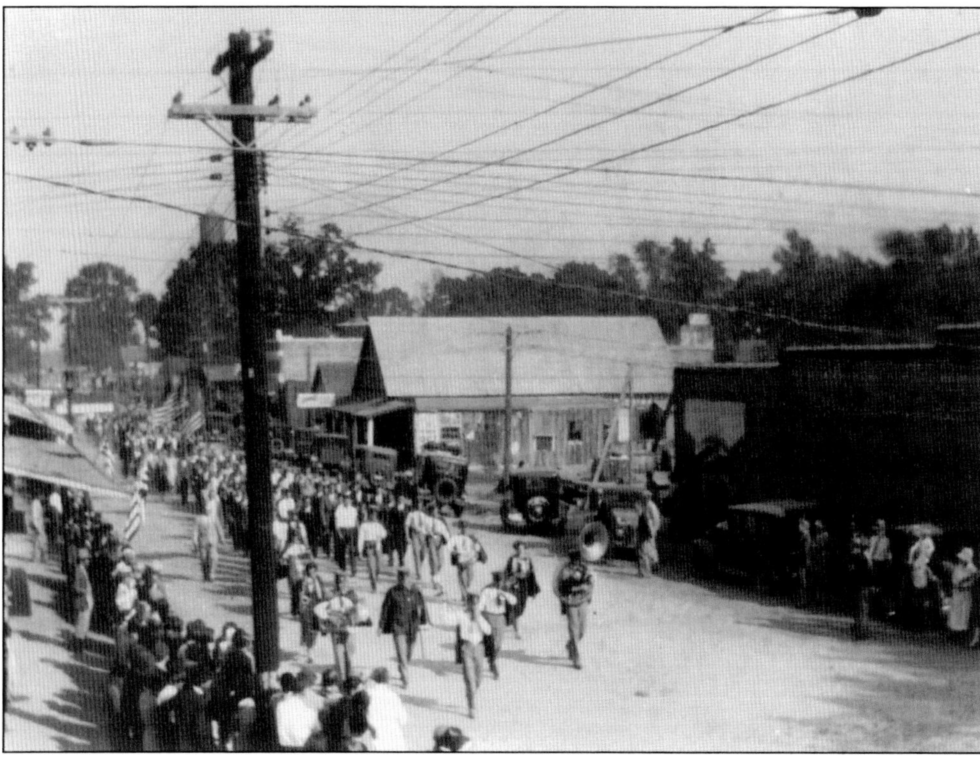

The Sheridan town band provides music for an Armistice parade on Oak Street celebrating the end of World War I. (Courtesy of Grant County Museum.)

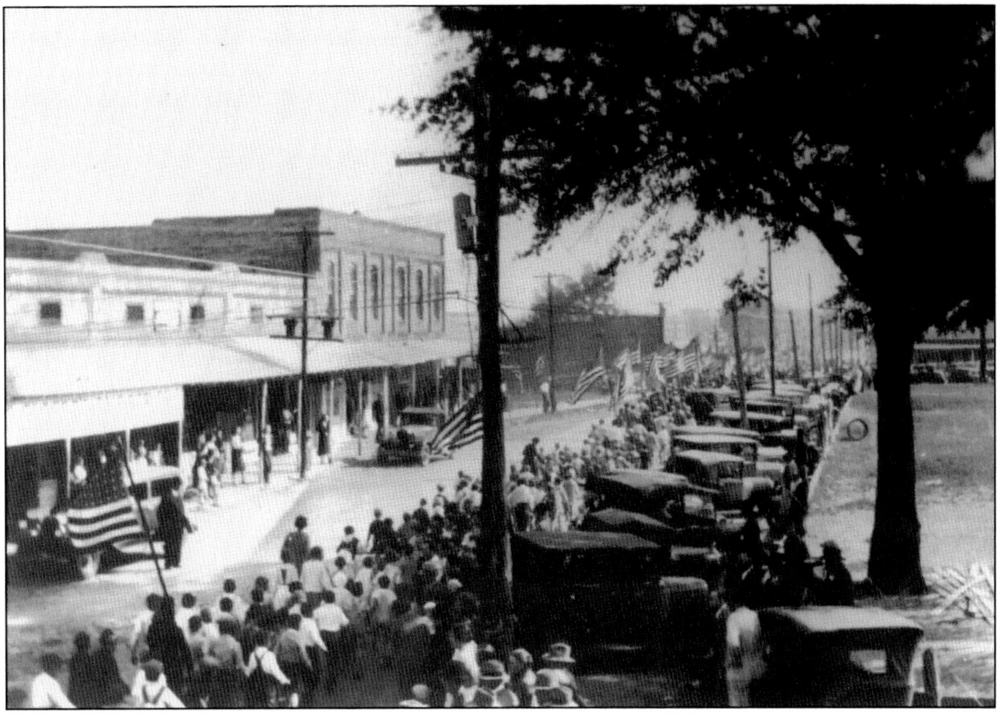

Lots of students and Model Ts appear on Center Street in front of the Grant County Courthouse as Sheridan celebrates Armistice Day with a parade. (Courtesy of Grant County Museum.)

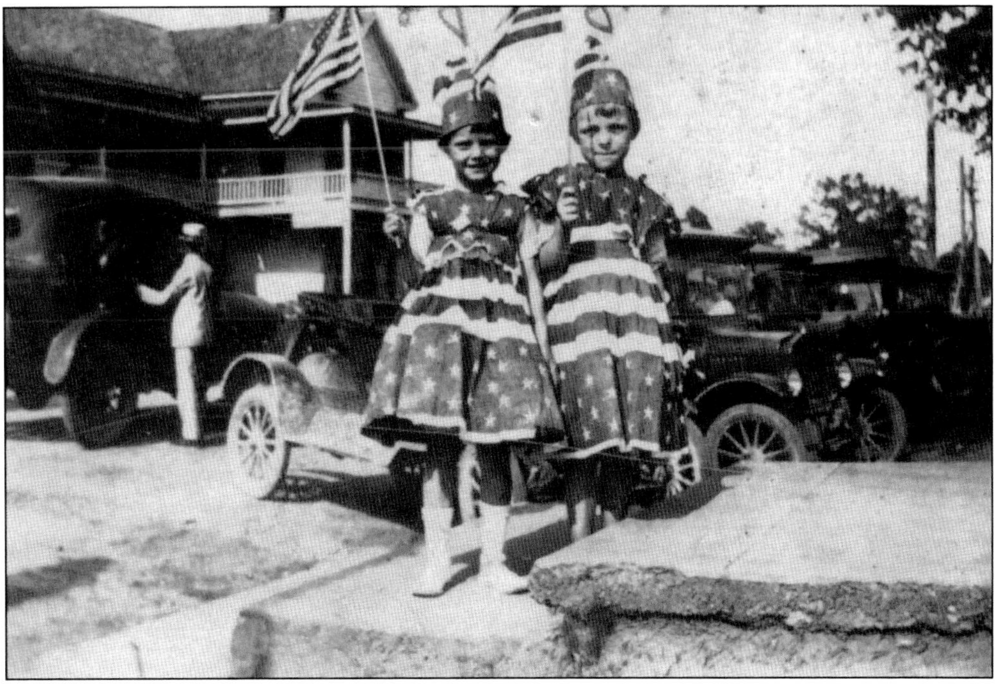

"Fourth day of July, 1919, near Courthouse" was written on the back of this patriotic photograph. Shown here are Retha Jones (right) and her friend Margaret. Retha's father, Lem Jones, operated a store on Center Street from the early 1900s until the 1940s. (Courtesy of Doug Lloyd.)

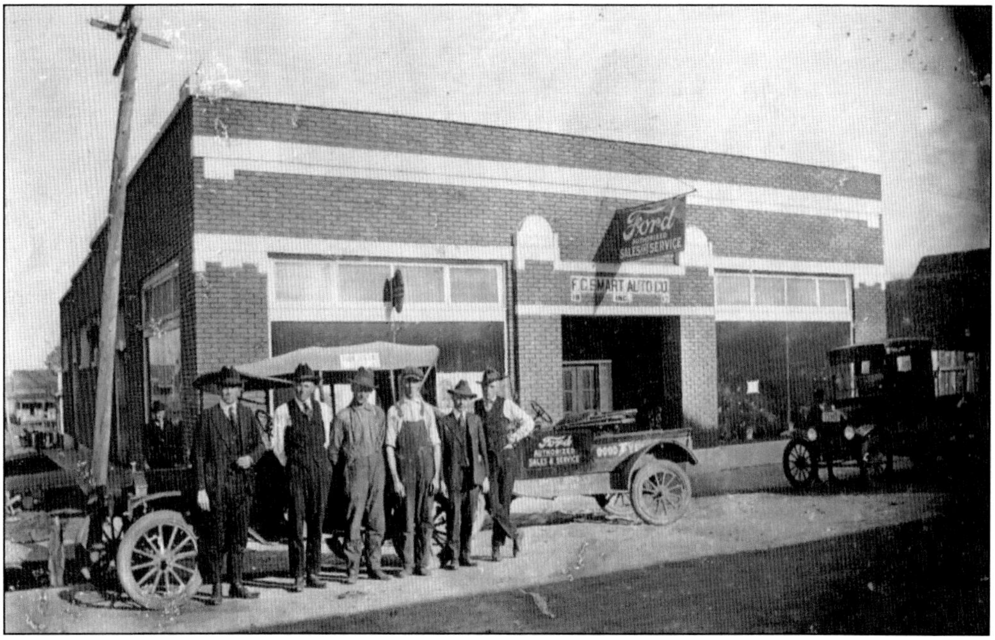


The Smart Ford Company dealt in Model Ts at the southeast corner of Oak and Bell Streets in Sheridan. The building would later catch fire, causing employees to pick up the cars and hand them out the front window to safety. The concrete foundation of the front of the building remained in 2012. (Courtesy of Grant County Museum.)

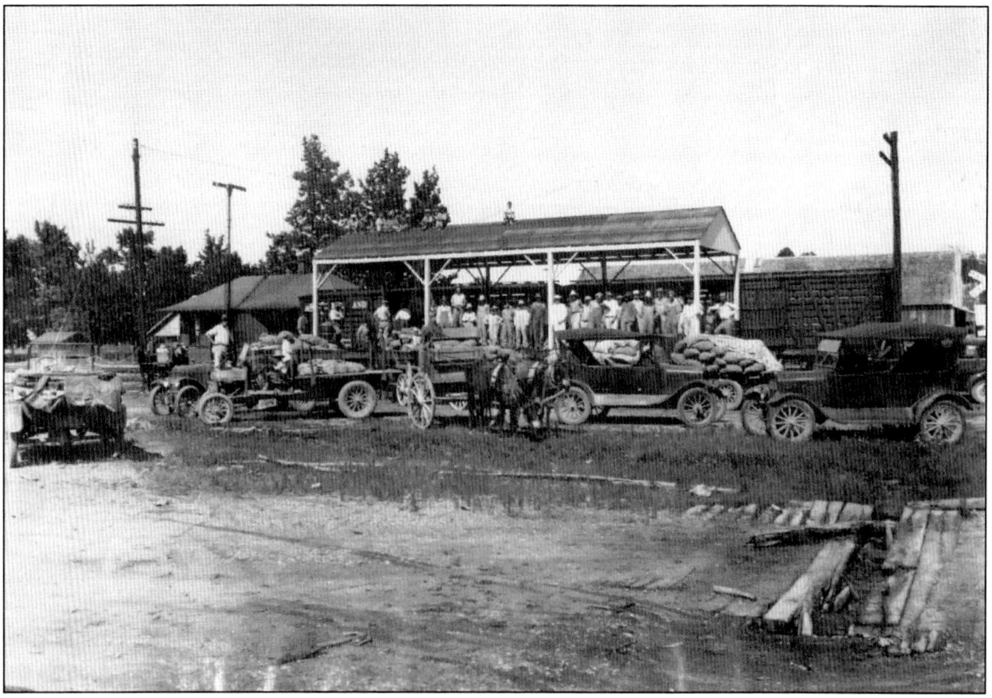

Poyen, in western Grant County, was once well known as a potato-growing area. In this 1920s photograph, farmers gather at the railroad station to ship the efforts of their labor. (Courtesy of W.H. Byrd Jr. family.)

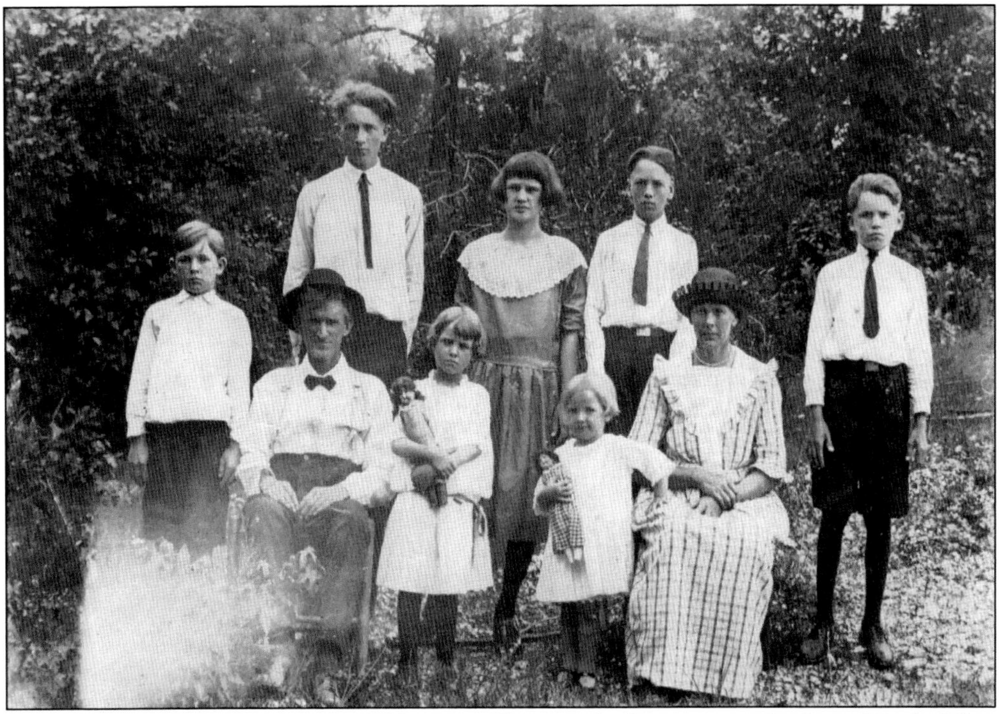

Rich and Lizzie Allen and their children are photographed in 1924 at the foot of Rich Allen Hill, west of Shiloh Church. Shown here are, from left to right, (first row) Thomas Boyce Allen, Richard Allen, Hazel W. Allen, Connie Cornelia Allen (the dolls were a gift from their "Grandpa Sam"), Lizzie Bratton Allen, and Jewel Wright Allen; (second row) Ivan Dean Allen, Wilma Altus Allen, and Robert Doyle Allen. (Courtesy of the author.)

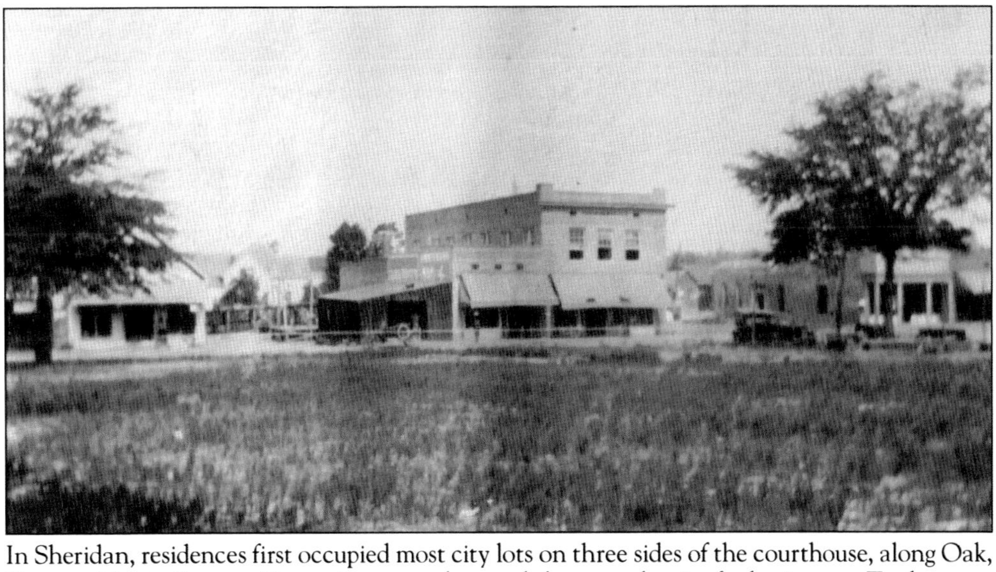

In Sheridan, residences first occupied most city lots on three sides of the courthouse, along Oak, Pine, and Main Streets. Center Street to the north became the site for businesses. To the west, on Oak Street, as seen in this 1920s photograph, was a garage for Model Ts. It stood south of the Bradley-Rushing Building (center) and the Grant County Bank (right). (Courtesy of Grant County Museum.)

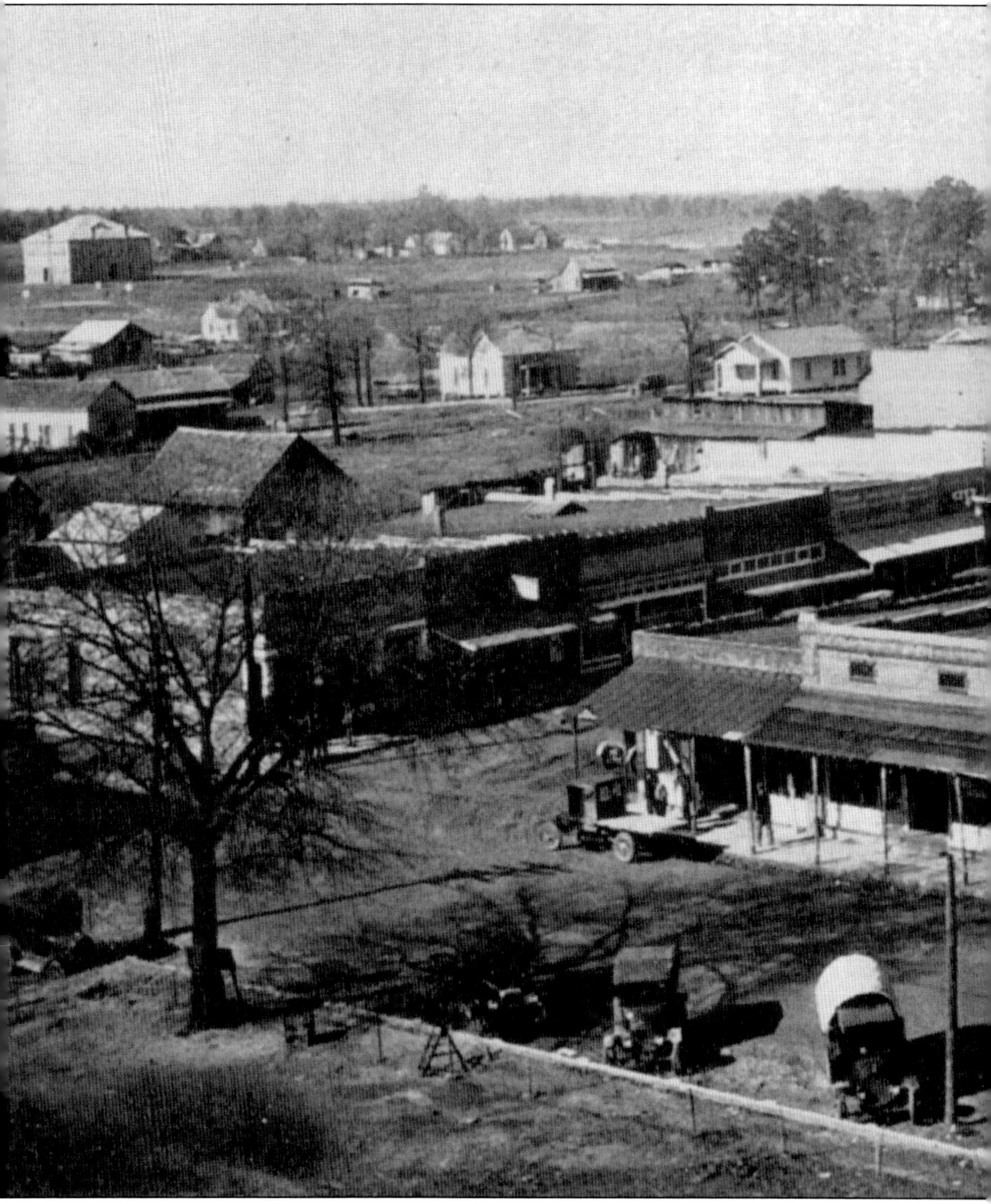

Taken from the Grant County Courthouse roof in 1926 is this photograph of Center Street (foreground) and Oak Street (extending toward top right). At far left is the Missionary Baptist College, just south of the current Sheridan High School campus. Around the corner on Oak Street stand wooden frames used to weigh and grade cotton bales. Further up Oak Street on the left are Shepherd's silent movie theater, south of, from left to right, the Blakely residence (still standing),

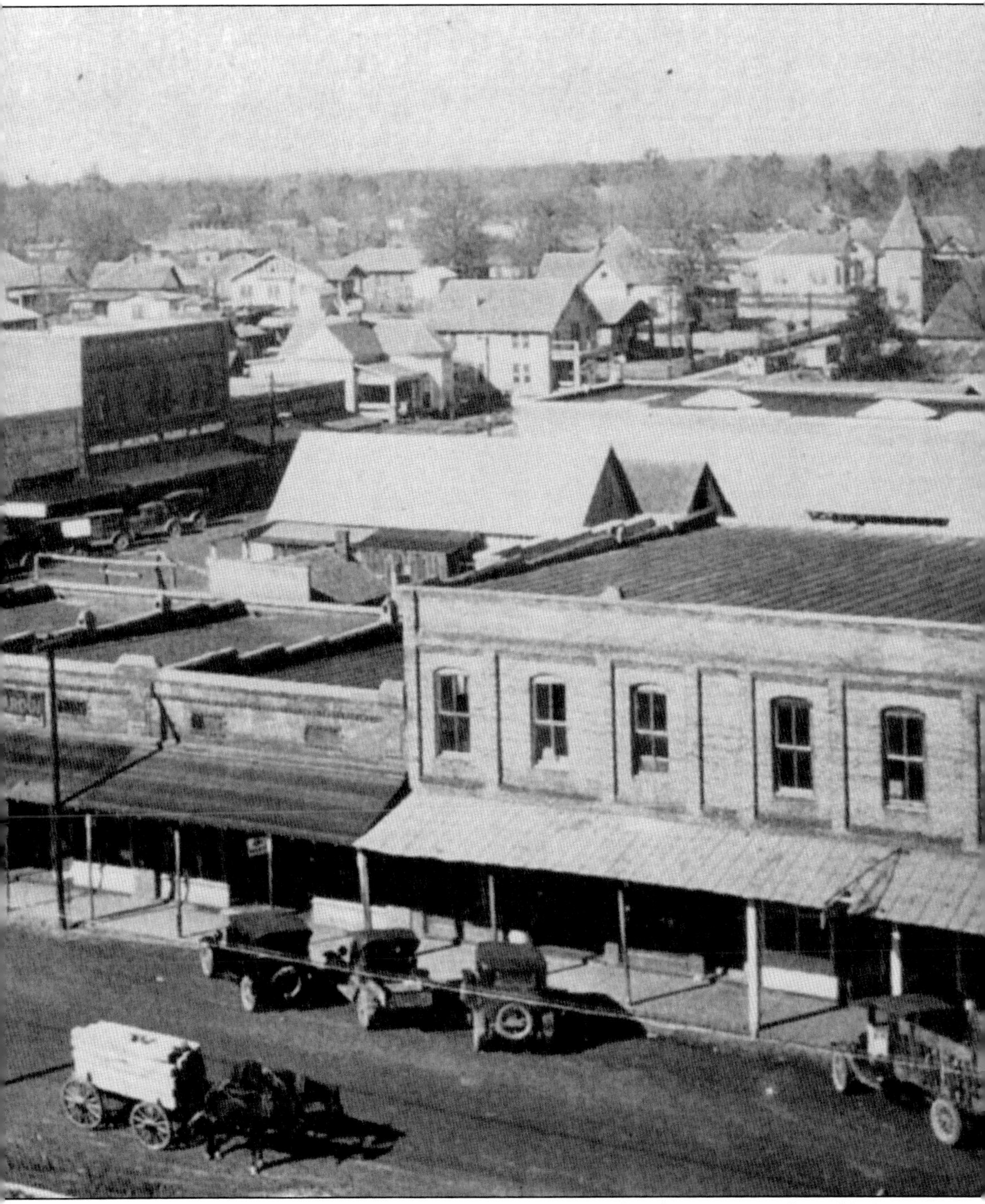

Aunt Bea Black's two-story boardinghouse (current driveway of Malvern National Bank), Oris and Gertrude Rushing's home, and Judge John W. and Della Lybrand's home (present-day bank parking lot). J. Dee Lybrand grew up at this home on the southwest corner of Oak and Church Streets. At top right is the tower of the Presbyterian church. (Courtesy of Sheridan School District yearbook.)

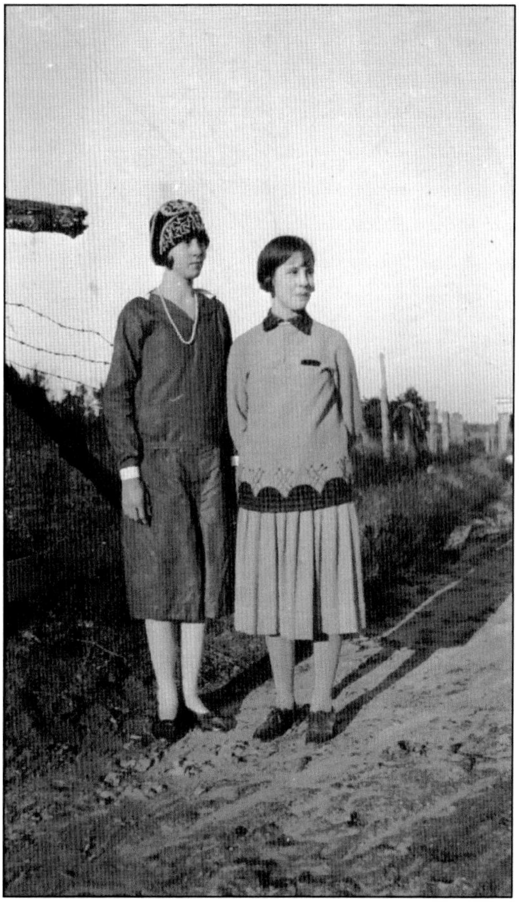

Posing alongside a street in Tull in the 1920s are Estelle DuVall Zahringer (left) and Edith DuVall Tull. (Courtesy of Joan Crowson Burton.)

A 1931 dance card invited guests to the Wilson Hotel, a center of social life in Sheridan for several decades. The hotel was located at the southeast corner of the intersection of highways 167 and 270. (Courtesy of Doug Lloyd.)

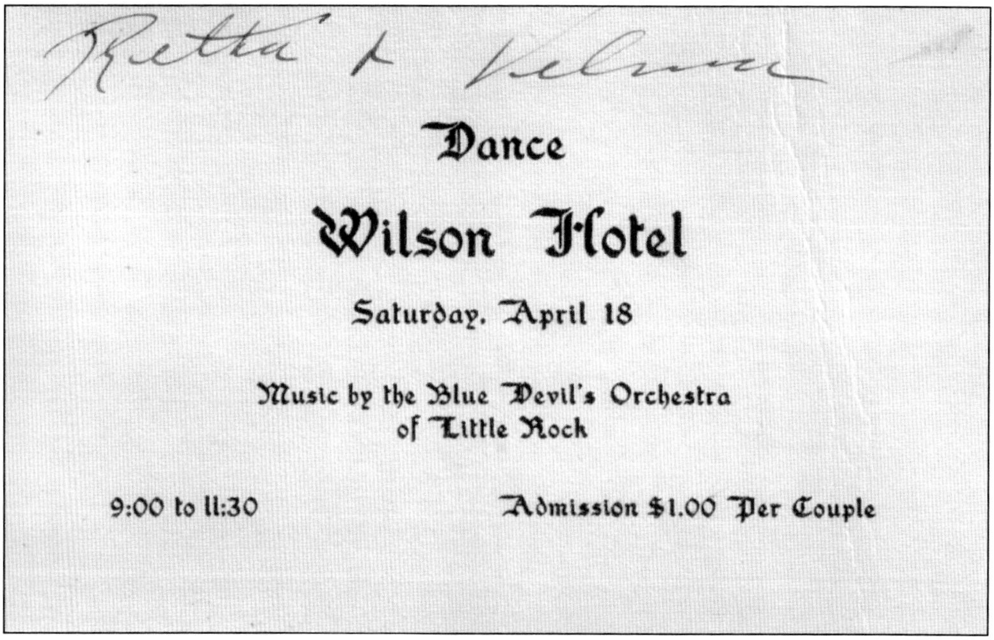

Retta & Velma

Dance

Wilson Hotel

Saturday, April 18

Music by the Blue Devil's Orchestra
of Little Rock

9:00 to 11:30 Admission $1.00 Per Couple

Sheridan residents Odie McCool (left), unidentified (center), and Mary Cabiness sport Sheridan sweaters. An unidentified escort stands in the background. (Courtesy of Grant County Museum.)

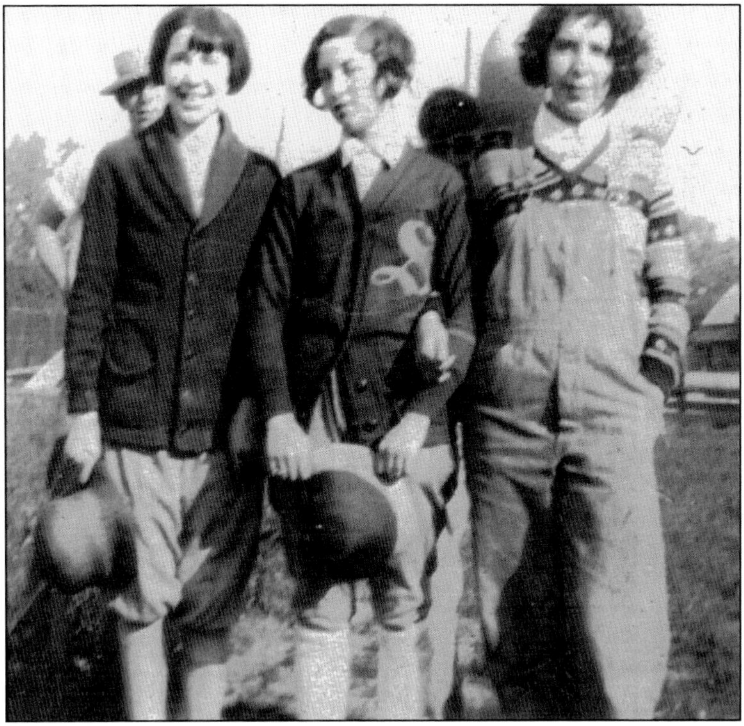

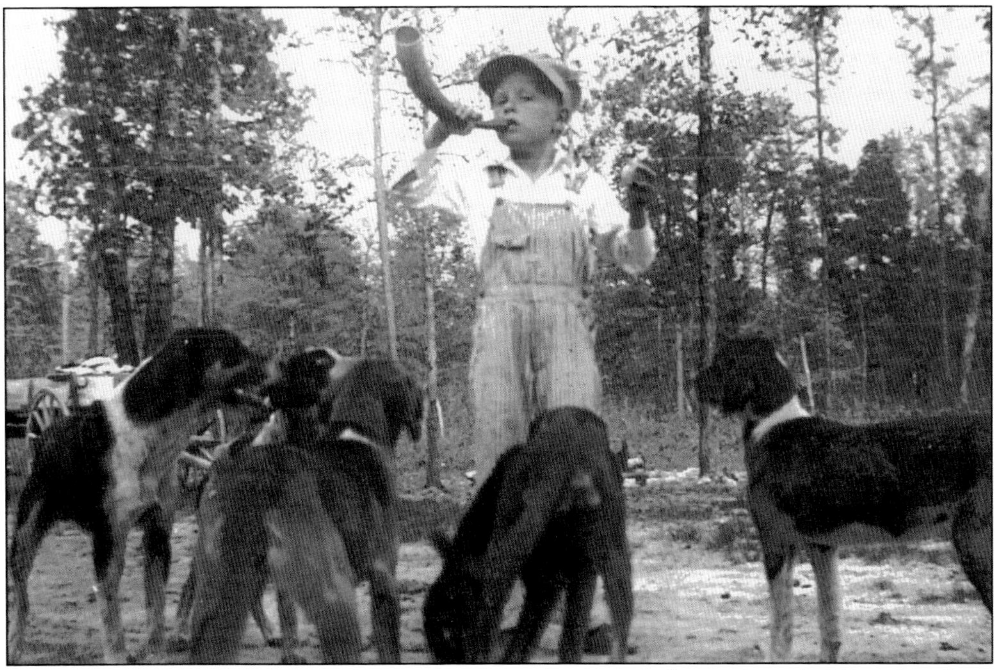

Fox hunting was still a very popular sport in Grant County in the 1920s and 1930s, when large numbers of people would attend annual Fox Hunter Association picnics. In this photograph, a youngster, L.D. Allen, born in 1929 to Ivan and Gracie Lou Stout Allen, attempts to call the dogs with a real horn. (Courtesy of Hazel Wilson.)

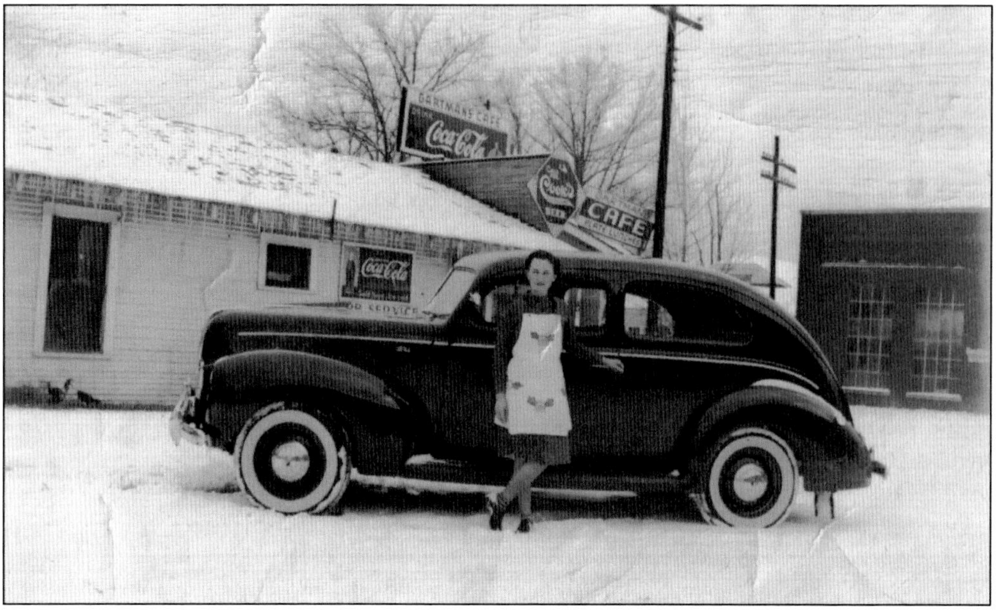


Mazel Pitts worked at Gartman's Café in Sheridan in the 1930s. (Courtesy of Grant County Museum.)

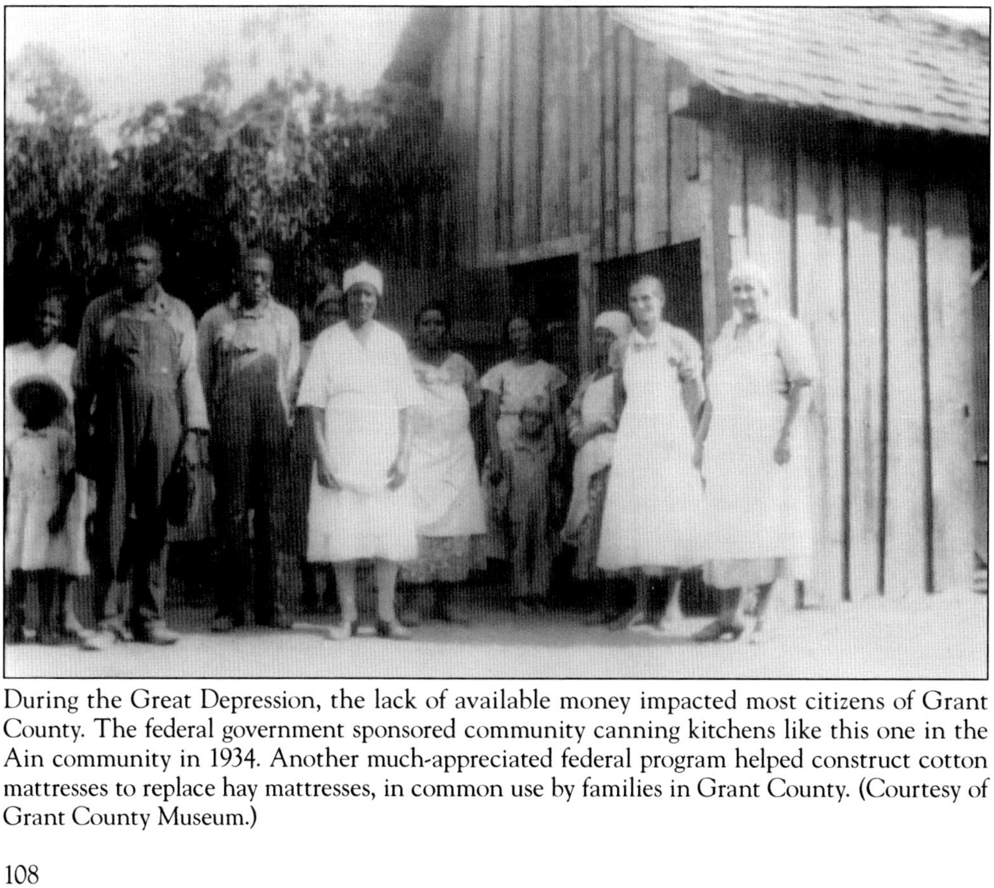

During the Great Depression, the lack of available money impacted most citizens of Grant County. The federal government sponsored community canning kitchens like this one in the Ain community in 1934. Another much-appreciated federal program helped construct cotton mattresses to replace hay mattresses, in common use by families in Grant County. (Courtesy of Grant County Museum.)

Few present-day residents of Grant County understand the hard physical labor it once took to chop cotton, lay it by, and pick the crop each fall in 20- to 40-acre fields. Pictured here in a field are, from left to right, Lizzie and Rich Allen, Connie Allen, Garlin Stout, and Hazel Allen in the background. (Courtesy of Hazel Wilson.)

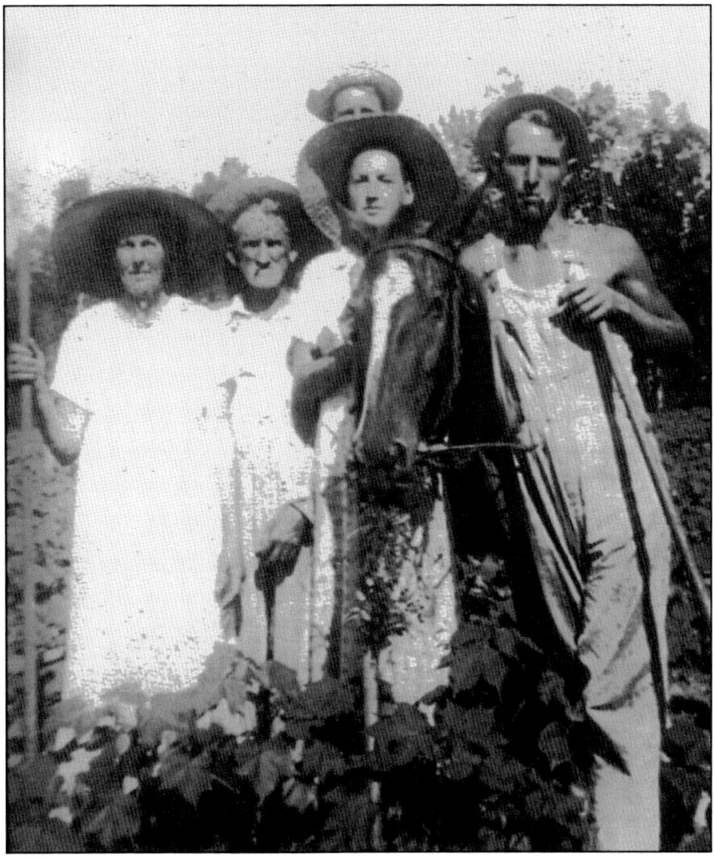

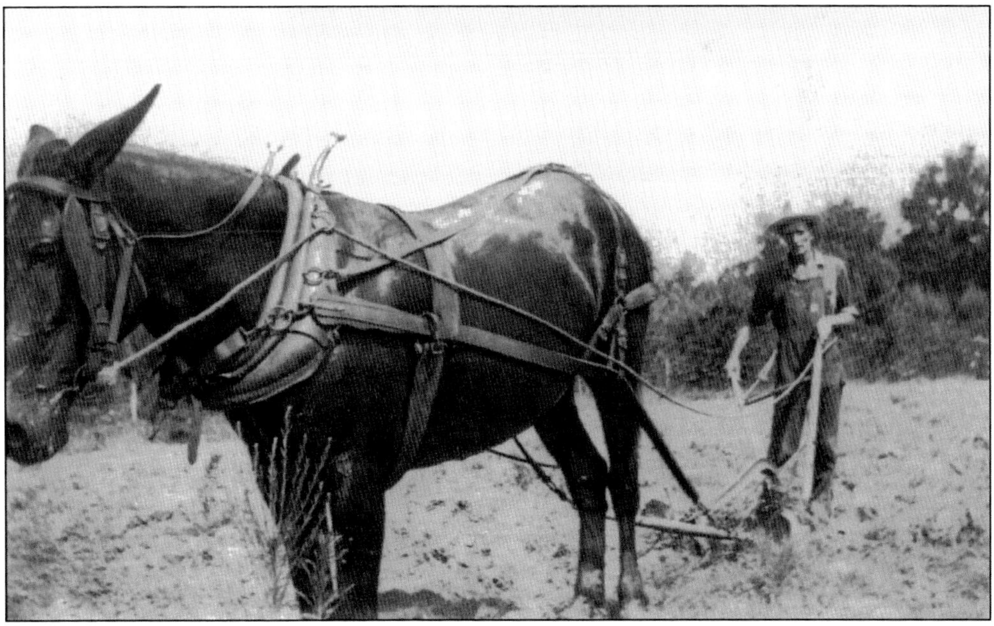

Plowing behind a horse in the 1930s was grueling work, as illustrated here by the tired look of the horse and the farmer, Ivan Allen, of southern Grant County. (Courtesy of Hazel Wilson.)

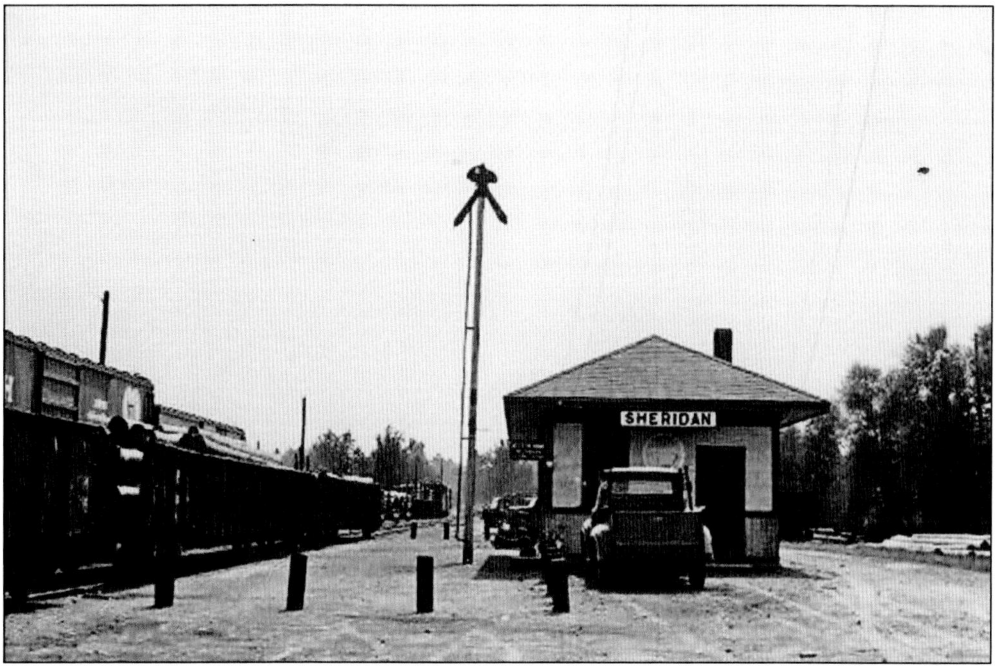


Sheridan once had daily passenger train service on the Choctaw Line of the PB&W Railroad. The depot was located along the tracks in the eastern part of the city, not far from the home place of Littleton Veazey, who donated land to create Sheridan as the county seat. This photograph was taken in September 1960 by *Arkansas Gazette* columnist Ernie Deane. (Courtesy of Arkansas History Commission.)

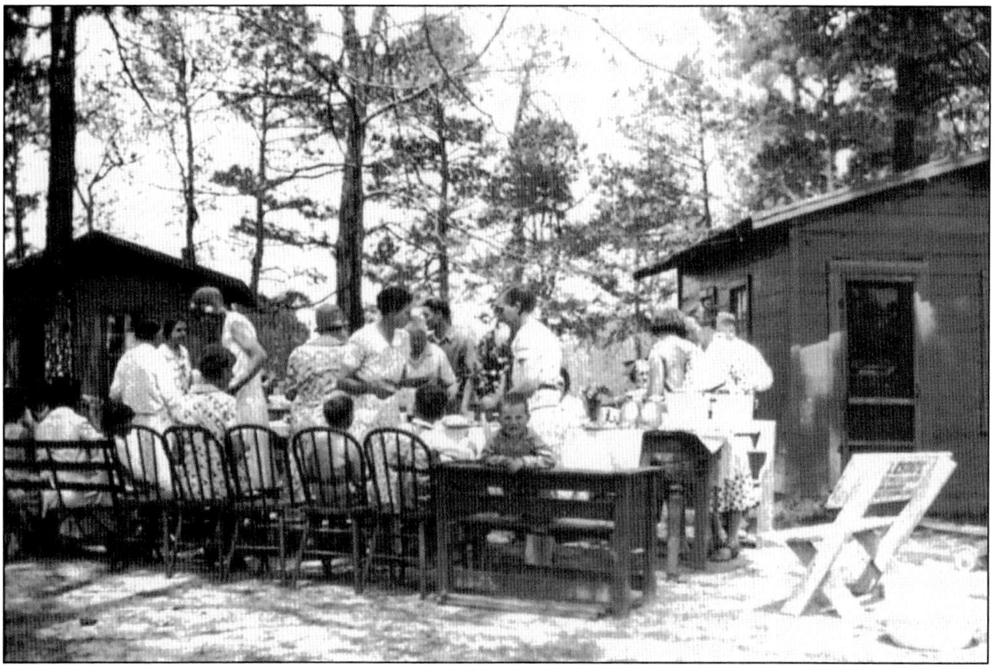

A tourist court located north of Sheridan hosts a meal in the 1930s. (Courtesy of Grant County Museum.)

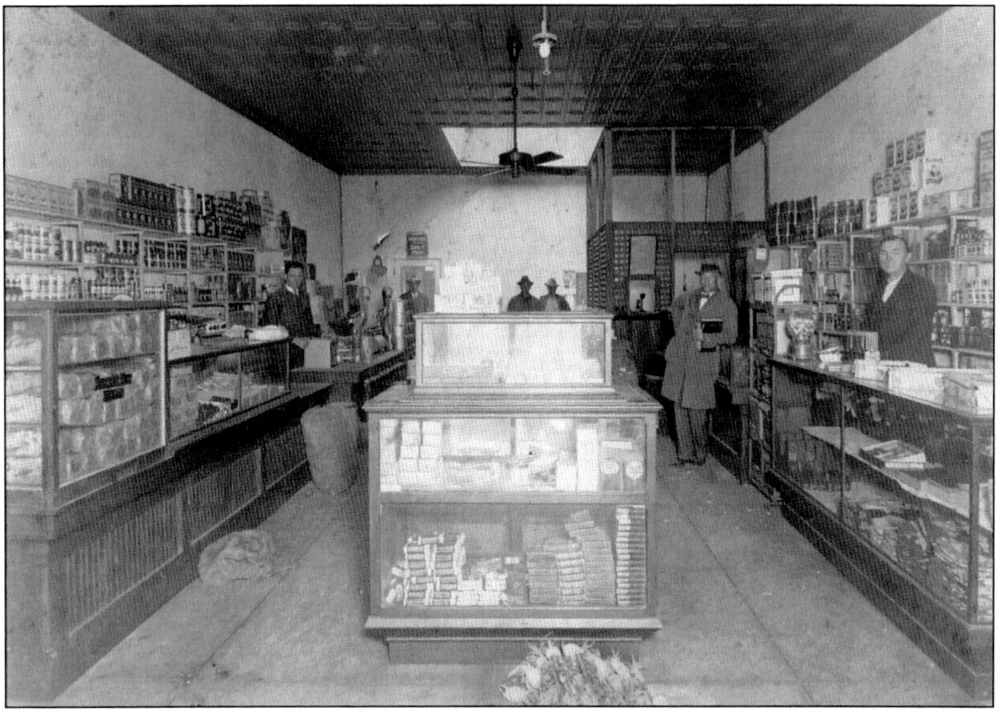

Lem Jones, at right, operated a general store at the northeast corner of Center and Main Streets for many years. He was the first to haul cotton bales by Model T from Grant County to Pine Bluff. During the Great Depression, he would buy produce from county residents to help their finances. Interested in progress, he once worked at the Grant County Bank, and he served on the school board for several years. (Courtesy of Doug Lloyd.)

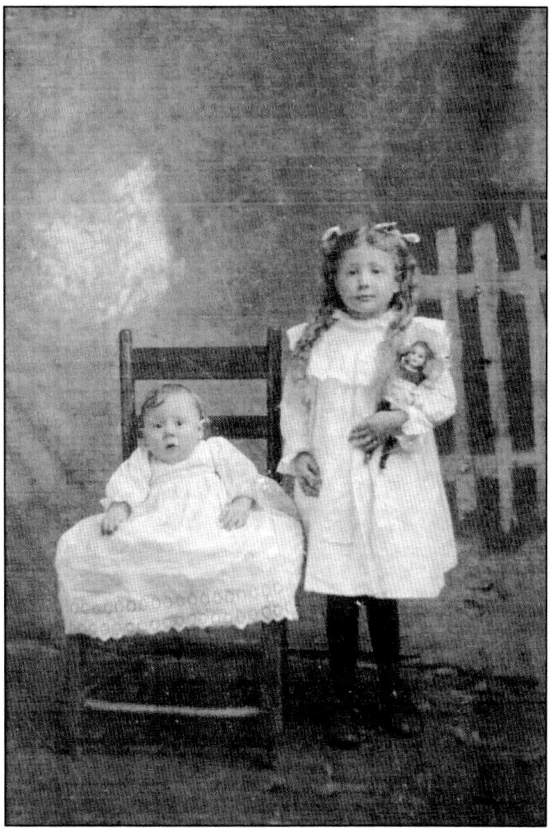

Shown here are Ode (left) and Lillian Brewer, children of Pink and Lavada Bratton Brewer of the Shiloh community. (Courtesy of Hazel Wilson.)

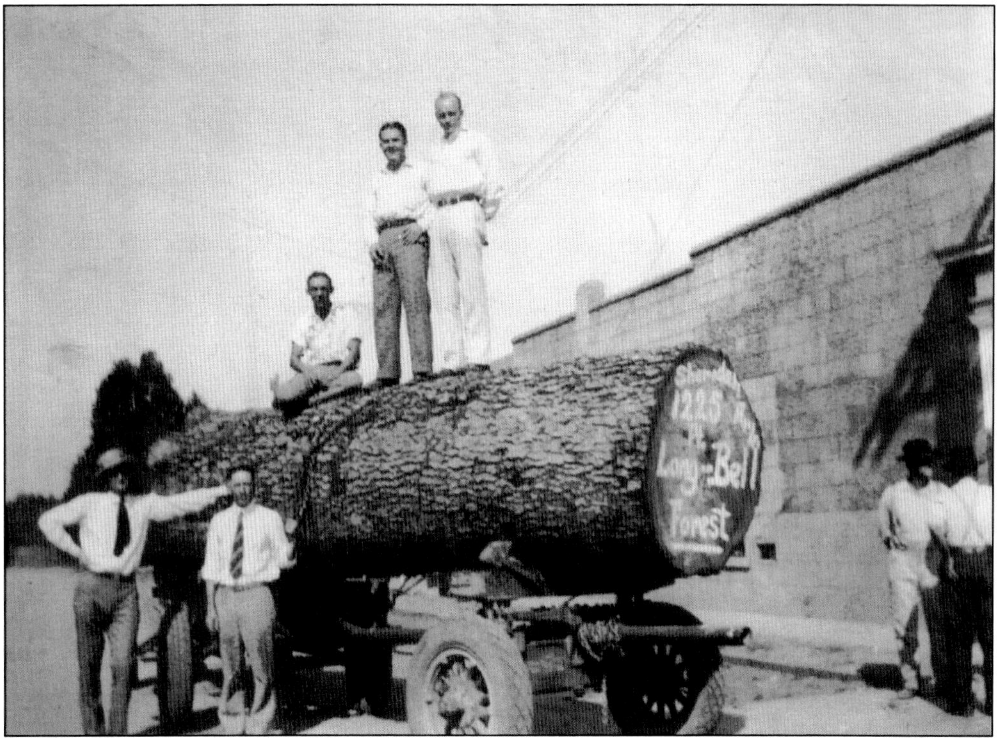


This log, containing 1,225 board feet, is proudly displayed by employees of the Long Bell Lumber Company of Sheridan. The company would provide income for many county residents. (Courtesy of Grant County Museum.)

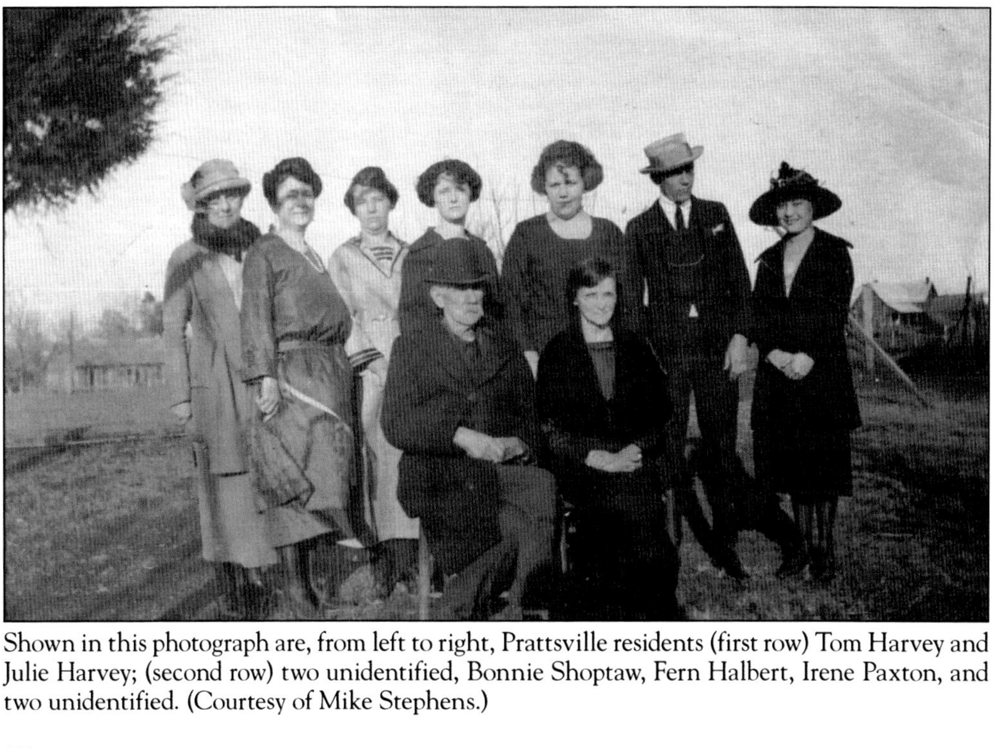

Shown in this photograph are, from left to right, Prattsville residents (first row) Tom Harvey and Julie Harvey; (second row) two unidentified, Bonnie Shoptaw, Fern Halbert, Irene Paxton, and two unidentified. (Courtesy of Mike Stephens.)

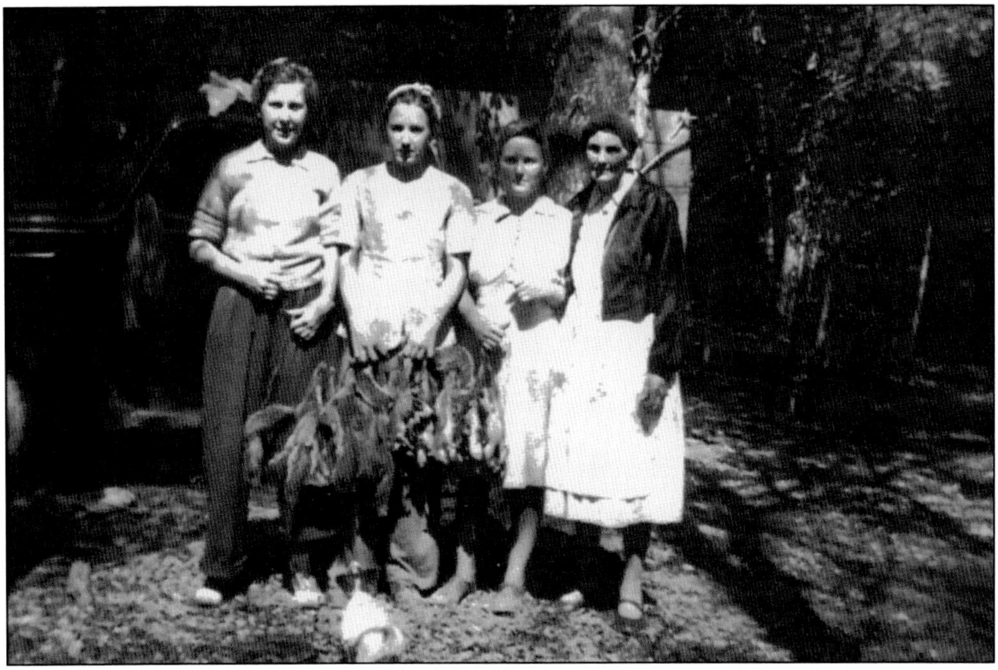

In the 1930s and 1940s, camping in the fall to hunt squirrels provided food for families during the harshest times of the Great Depression and of rationing during World War II. Shown here in the Saline River bottoms in October 1943 are, from left to right, Elsie Moring Allen, Connie Allen, Hazel Allen Wilson, and Lizzie Bratton Allen. (Courtesy of Hazel Wilson.)

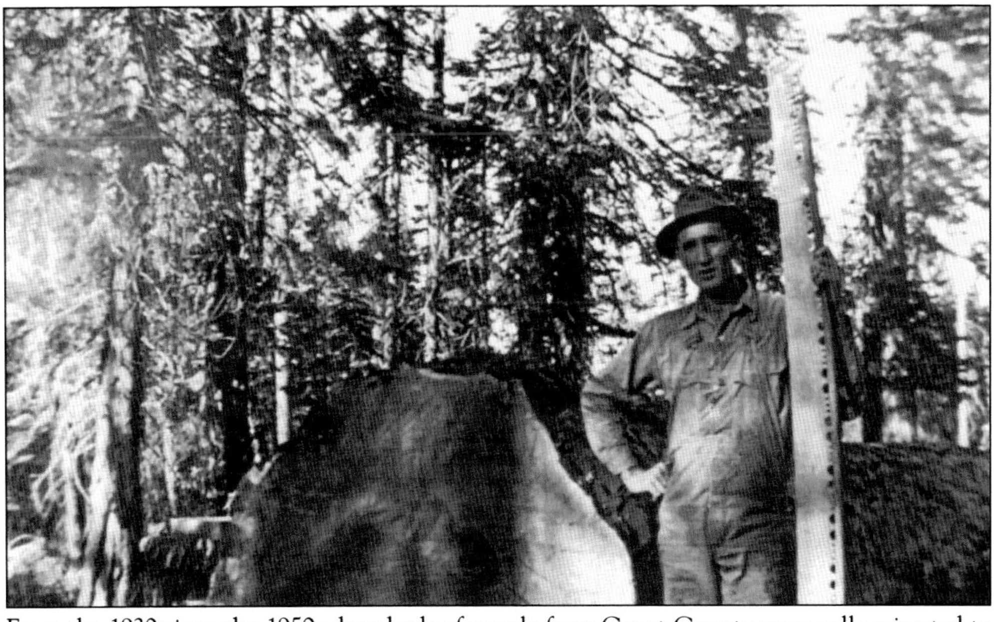

From the 1930s into the 1950s, hundreds of people from Grant County seasonally migrated to northern California and southern Oregon in search of timber jobs. Many decided to stay. In this photograph, Ivan Allen uses a crosscut saw on one such journey in California. As a result of harsh economics, the population of Sheridan fell from 1,590 in 1930 to 1,338 in 1940, a decline of 252. (Courtesy of Hazel Wilson.)

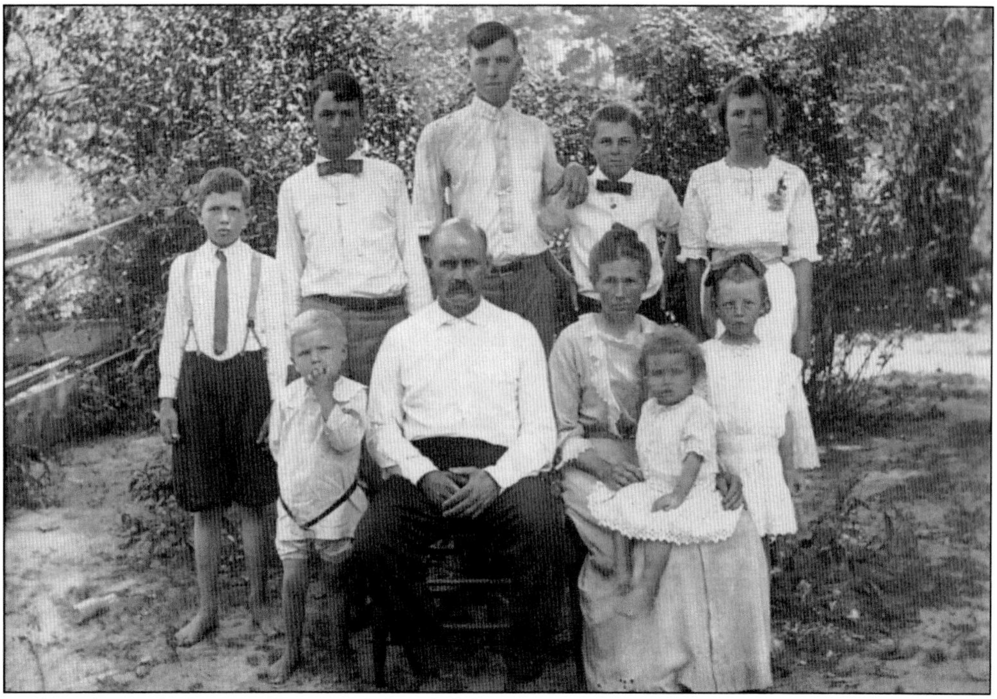

Pink and Lavada Bratton Brewer of the Shiloh community pose with their younger children in this photograph. Shown are, from left to right, the following: (first row) Scott, Pink, Lavada, Ruth (on lap), and Elsie; (second row) Otis, Earl, Girven, cousin Gus McMillan, and Lillian. Daughter Altus Brewer Rhodes is not pictured. (Courtesy of Hazel Wilson.)

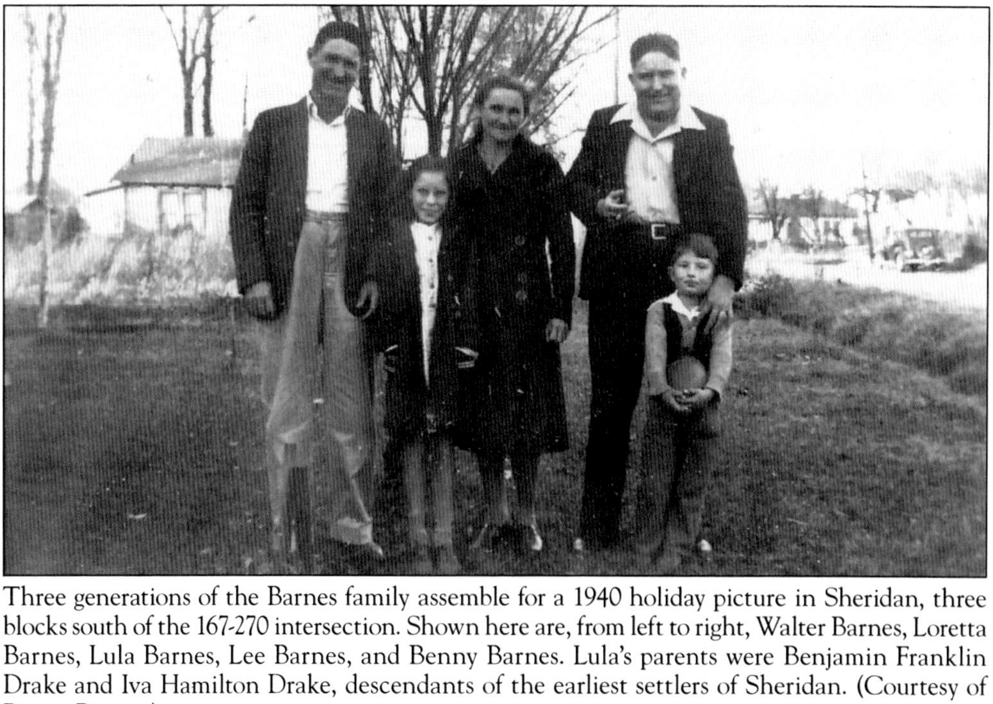

Three generations of the Barnes family assemble for a 1940 holiday picture in Sheridan, three blocks south of the 167-270 intersection. Shown here are, from left to right, Walter Barnes, Loretta Barnes, Lula Barnes, Lee Barnes, and Benny Barnes. Lula's parents were Benjamin Franklin Drake and Iva Hamilton Drake, descendants of the earliest settlers of Sheridan. (Courtesy of Benny Barnes.)

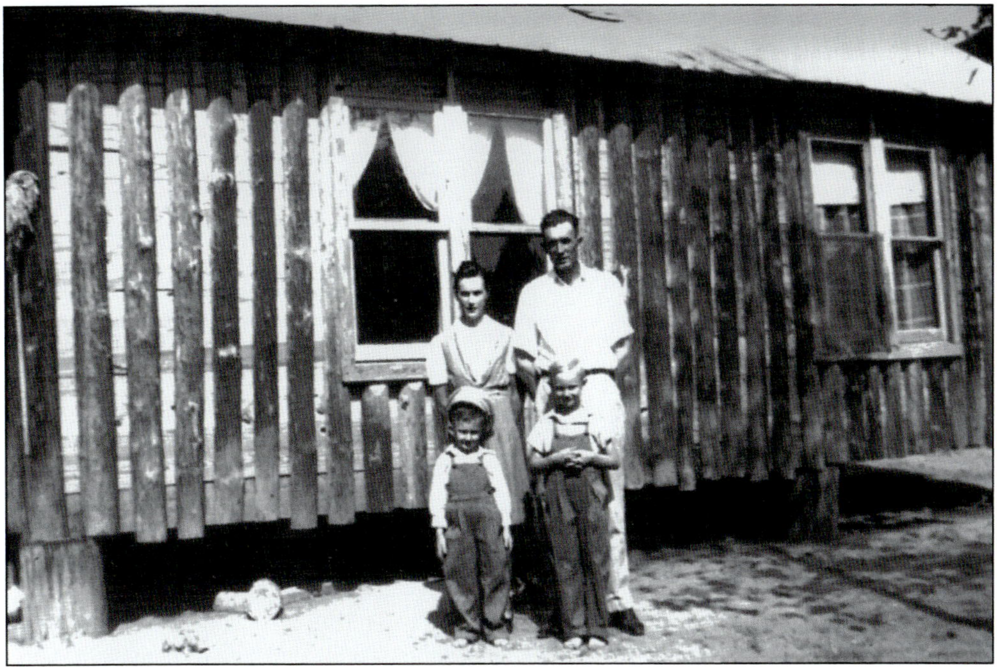

Ivan and Laura Allen and sons, Rommie (left) and Bob, pose in 1940 at their home north of Shiloh Church. (Courtesy of Hazel Wilson.)

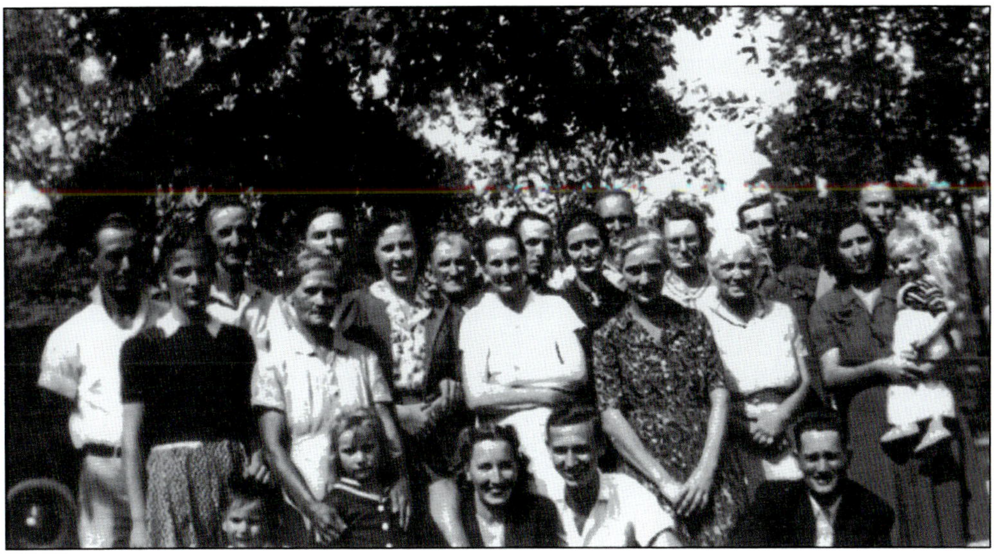

A family reunion and meal was held on October 3, 1943, as a sendoff for Marvin Allen, who was joining the US Army in World War II. Shown here are, from left to right, (first row) Fred Allen, Sammy Allen (children of Tootie and Valerie Allen), Connie Allen, Marvin Allen, and Davis Wilson; (second row) Valerie Allen, Rose Walker Allen, Elsie Moring Allen, Lizzie Bratton Allen, Tennie Ward Wilson, Sally Allen Ward, and Clova Nell Allen Wilson (holding Roy Joe Wilson); (third row) Arnold "Tootie" Allen, "Shorty" Allen, Boyce Allen, Rich Allen, Buck Allen, Fannie Martindale Allen, Earl Barnes, Lois Hale Barnes, Charlie Ward, and Havis Wilson. Sgt. Marvin Allen, aged 19, arrived in England on April 20, 1944, was with the first troops to land on D-Day, and was killed in France on July 31. (Courtesy of the author.)

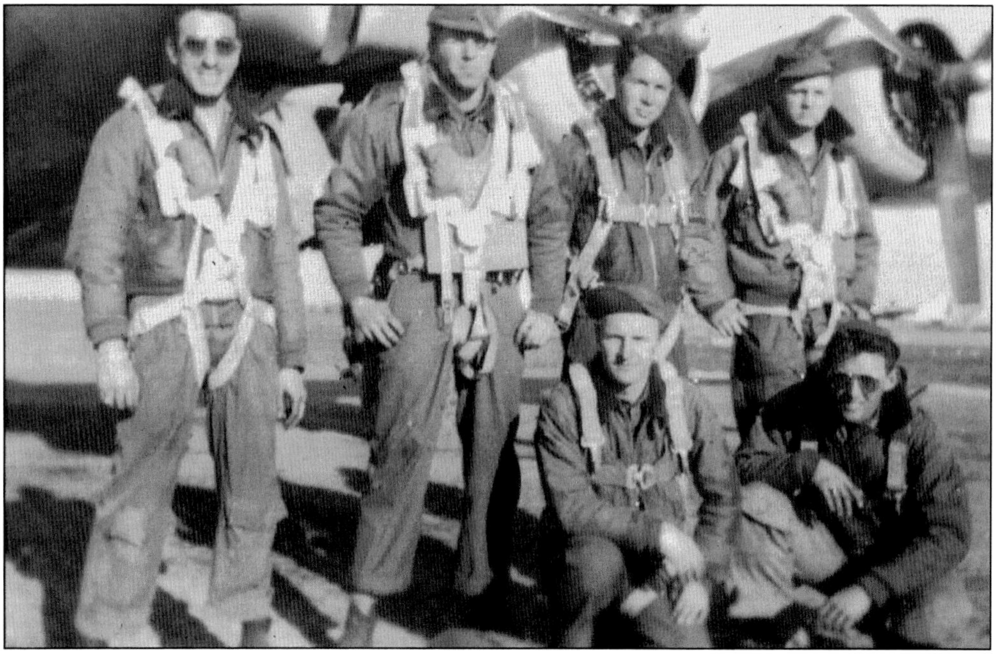

Dr. Curtis B. Clark of Sheridan survived 26 successful missions as a B-17 right waist gunner in the 483rd Bombardment Group, known as the Red Tails, based at Foggia, Italy, during World War II. Shown here are, from left to right, (first row) Paul U. Johnson (senior enlisted officer, top turret, from Boston, Massachusetts) and Calvin W. Holder (ball turret gunner, from Electra, Texas); (second row) Lou Garthe (left waist gunner, from New Jersey), Norman L. Branch (radio gunner, veteran of Pearl Harbor, from North Carolina), Frederick P. Boudouske (tail gunner, from New Orleans, Louisiana), and Curtis Clark (right waist gunner). They were covered during combat by the Tuskeegee Airmen. (Courtesy of Dr. Curtis B. Clark.)

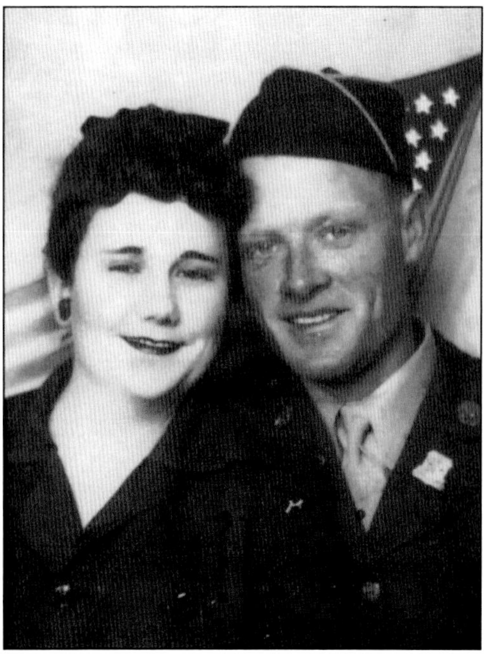

Okie Theron Reynolds is pictured with the love of his life, Gladys Fee, in December 1943. Okie T. survived being wounded during the Battle of Leyte Gulf in the Pacific during World War II. (Courtesy of Peggy Reynolds.)

Billy Hope of Sheridan, son of Dr. and Mrs. O.W. Hope, served in the US Navy during World War II. (Courtesy of Grant County Museum.)

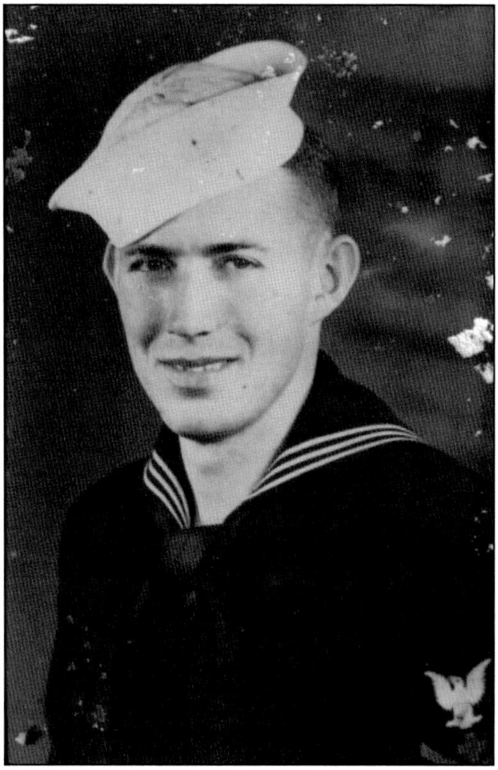

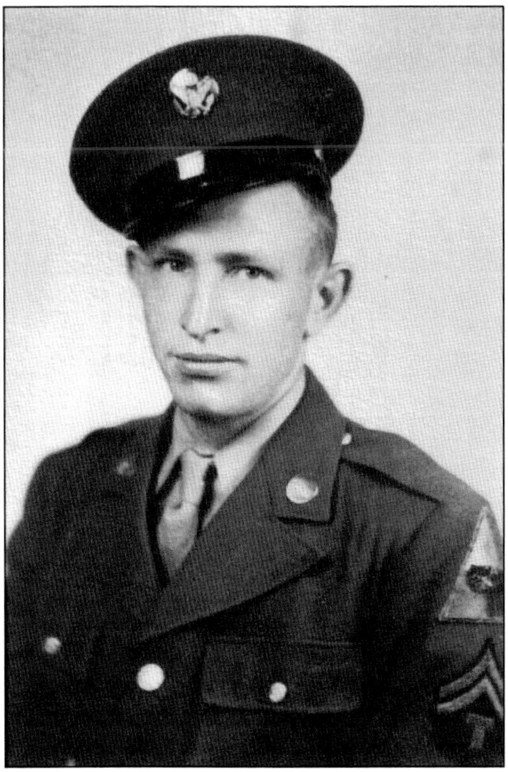

TSgt. Garlin H. Stout grew up in the Shiloh community and worked at J.L. Williams and Sons before entering the Army in 1942. After training with a tank unit in Virginia, he was based in England from January 1944 until D-Day. He was wounded in Germany on October 17 and died the next day. He left behind his wife, Cassie, and two little boys, Garlin Franklin (three years old) and Jerry Henry (one), the latter whom he never saw. Garlin had written home saying, "I'd give my right arm to see my little boy." (Courtesy of Grant County Museum.)

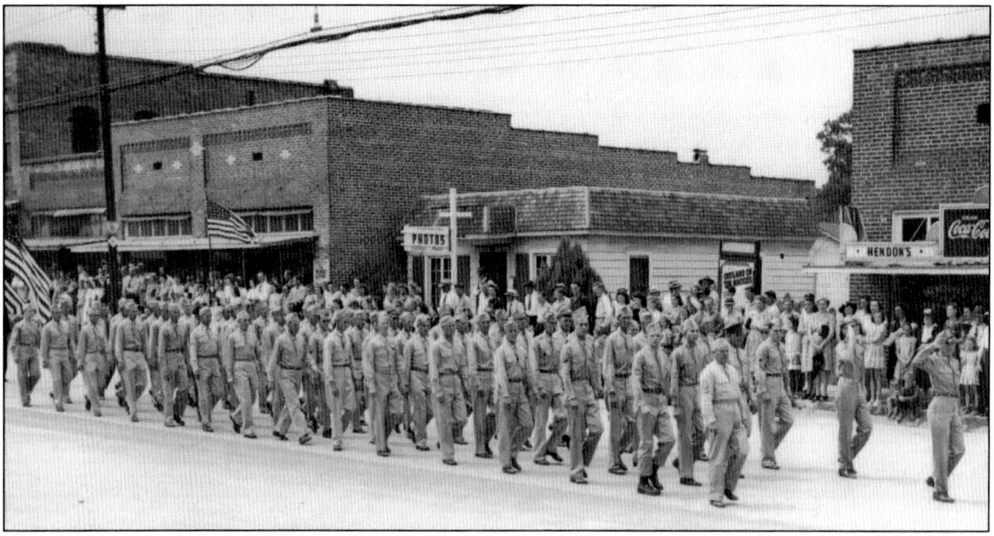


World War II veterans march east on Center Street before the 1946 dedication of the Grant County War Memorial, located on the courthouse lawn. The concrete street was built as a federal aid project in the 1930s. Standing under the 48-star American flag, left of the man with a tie, are, from left to right, Lois Lea Rushing Clark, Mary Frances McDonald Harper, and Helen Bratton. The tall marching soldier below the east window of Eifling Photos is Baptist minister Brother Joe May. The sixth soldier from the front in the near column is Calvin Perry, with his cap tilted left on his high forehead. (Courtesy of Doug Lloyd.)

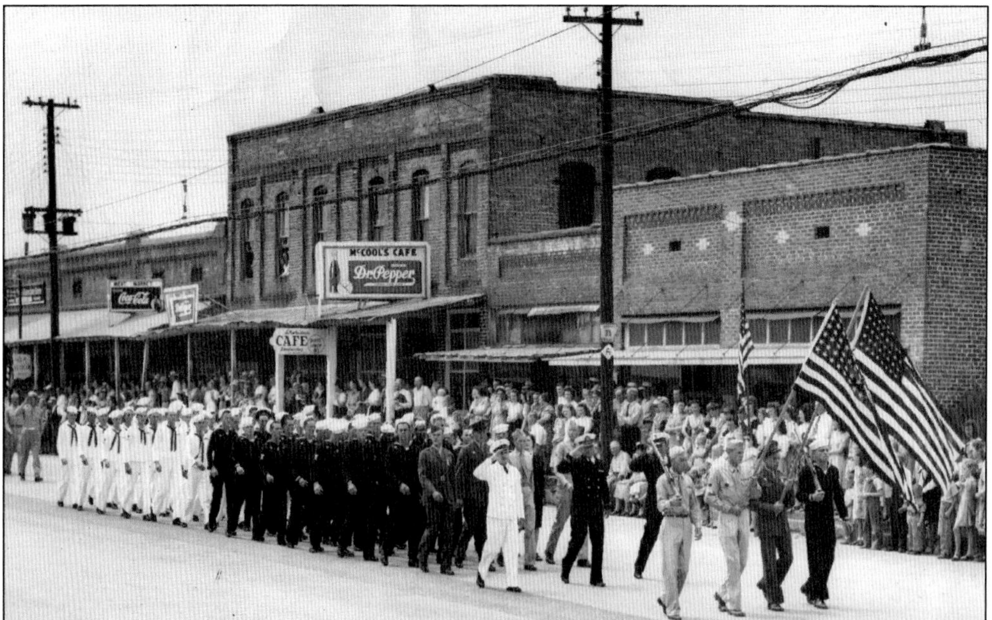


Local World War II US Navy veterans march on Center Street to dedicate the 1946 Grant County War Memorial. Standing on the sidewalk, directly above the left shoulder of the saluting soldier in the all-white uniform, is Mary Francis McDonald Harper; to her left, in profile, is Lois Lea Rushing Clark. Helen Bratton stands to Mary's right. Plate lunches were advertised for 40¢. (Courtesy of Doug Lloyd.)

Cowboys to the core, Jerry Humphries (left) and Mike Stephens are ready to ride the range in 1946 in Prattsville. (Courtesy of Mike Stephens.)

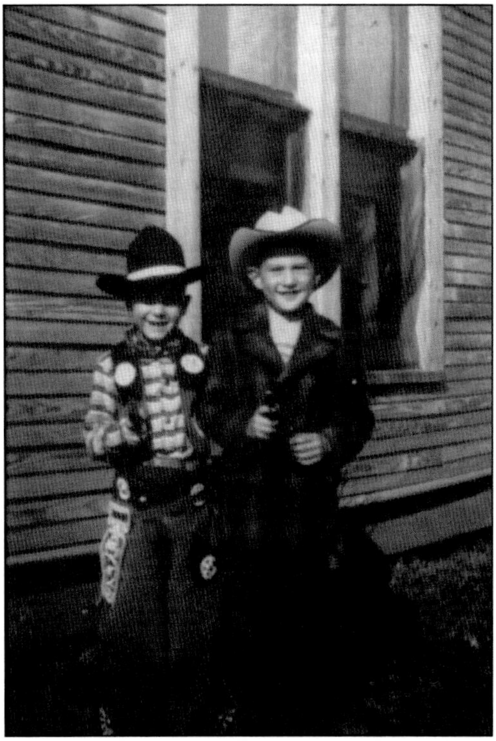

Leading women of Sheridan pose for a photograph. Shown here are, from left to right, (first row) Mrs. Lynn Nall, Clyde Nall (Juff), Rena Shepherd (Barney), Ruth Otts, Vera Blakely, Mrs. H.Q. Haley, and Minnie Nall (Tye); (second row) Delma Walker Rushing (Earl), unidentified, Hattie Nelson (Robert), Claudia Jones (Lem), Mrs. Newell, unidentified, Mrs. Slaughter, Mrs. Ed F. McDonald, Marjorie Goolsby (Hershel), and Muriel Nall (Dan). (Courtesy of Doug Lloyd.)

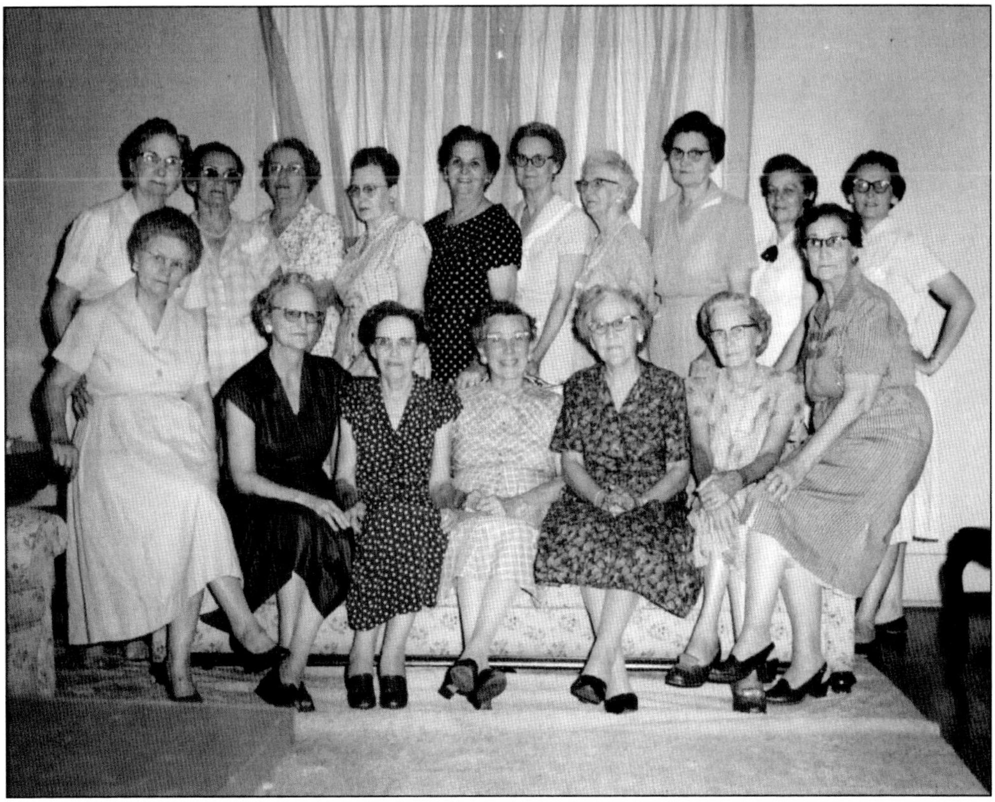

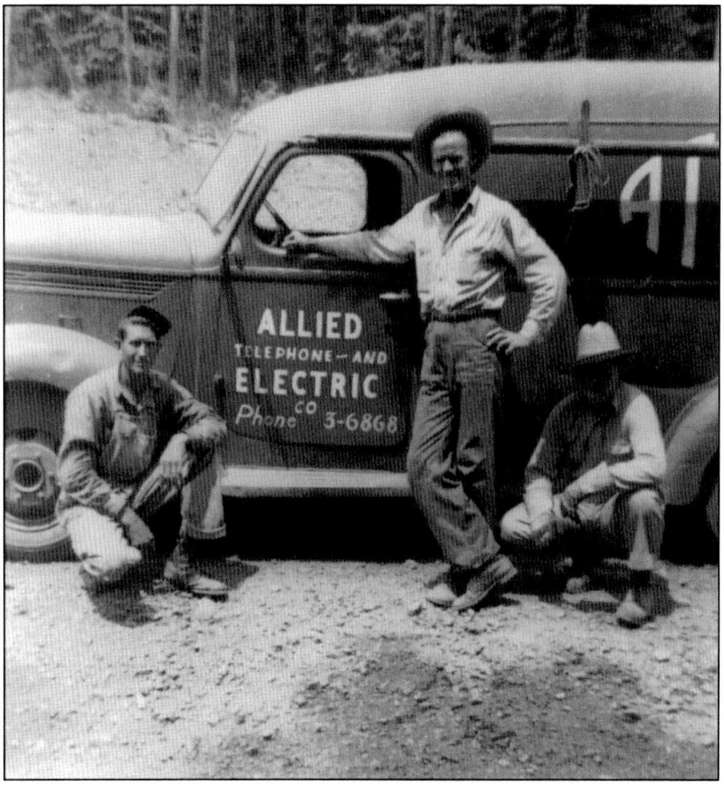


The first construction workers of Allied Telephone and Electric are pictured in September 1947 doing contracting work near the newly constructed Narrows Dam at Murfreesboro. Shown here are Grant County natives Carl McGarity (left) and Marvin (Boe) Wilson (center), with Ben Junior Brewer (right). (Courtesy of the author.)

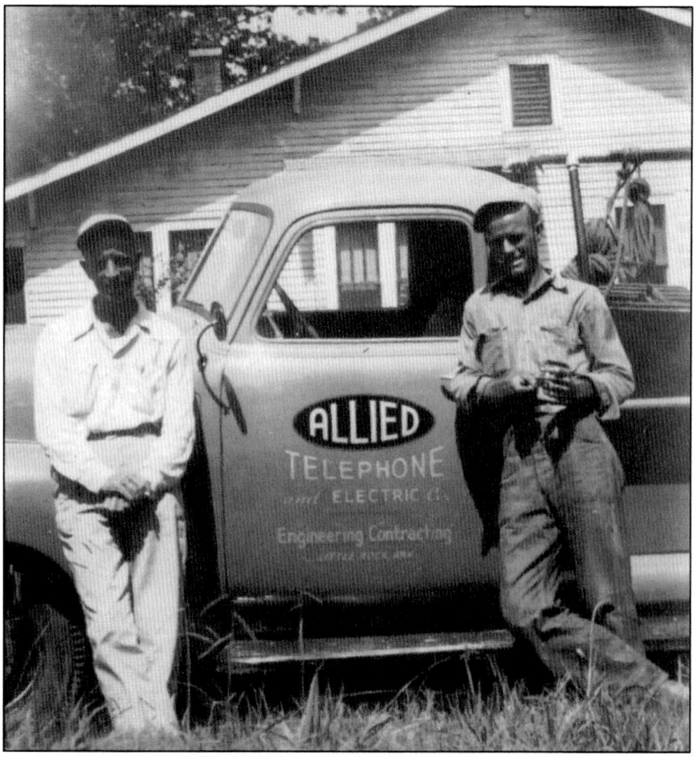


C.B. Miller (left), cofounder of Allied Telephone, and early employee Marvin (Boe) Wilson of Grant County, are shown at work in Stuttgart in 1949. On the photograph's reverse is written, "The engineering department at that time." Allied Telephone's first acquisition was the Grant County Phone Company of Sheridan. After years of steady growth under cofounder Hugh Wilbourn and Joe Ford, Allied became ALLTEL in 1983 before being divided in 2006 into Windstream and ALLTEL wireless, which was purchased in 2008 by Verizon. (Courtesy of the author.)

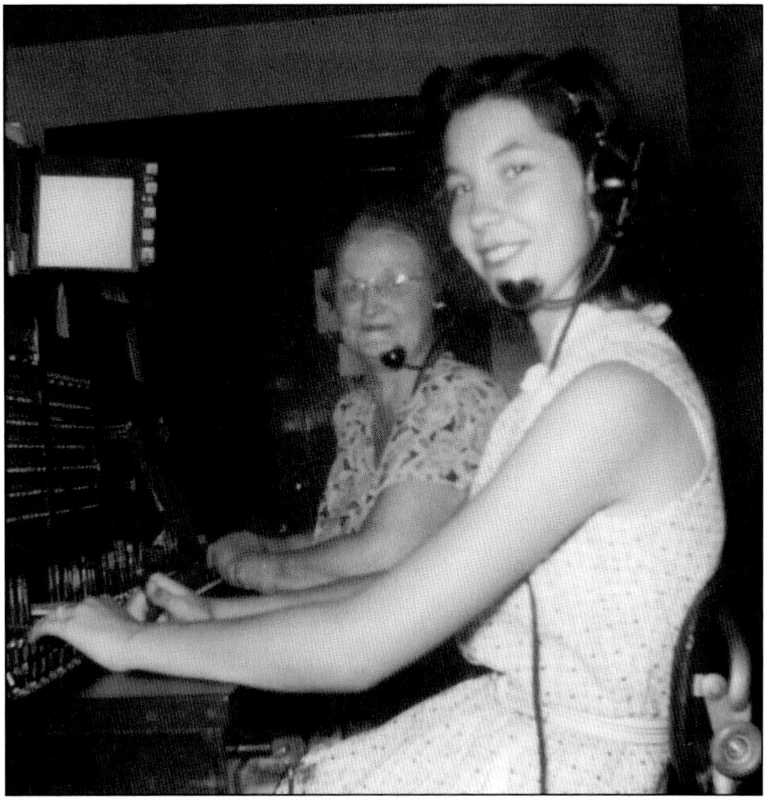

Grant County natives Pearl Hamilton (left) and Nellie Hamner Wagner served the public as switchboard operators at the phone company in Sheridan. When Allied Telephone began ownership of the company, wooden phones with cranks, 15-family party lines, and phone wires attached to trees were common in the county. (Courtesy of Grant County Museum.)

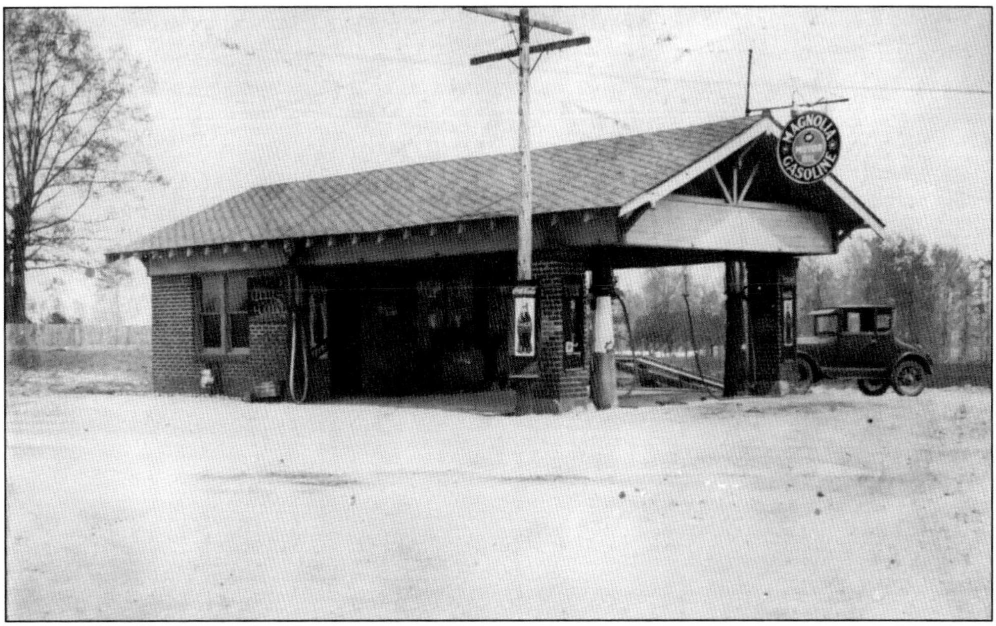


The Robert Clark service station occupied the northeast corner of the intersection of Vine Street and Highway 35 at a time when gasoline was 24¢ a gallon. Later, Avant's Grocery would occupy the site, now the location of Sheridan's First Assembly of God Church. (Courtesy of Grant County Museum.)

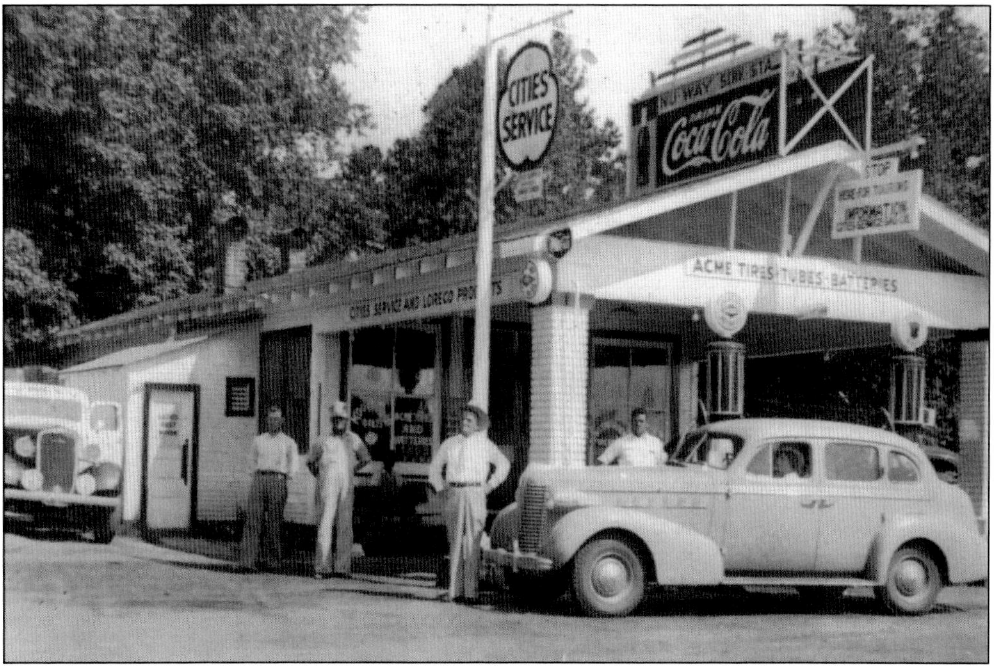


Located north of the Missouri Pacific railroad tracks, on the west side of Highway 167 north of Sheridan, was the Nu-Way Cities Service Station (just north of present-day Stockton Power Equipment). Pictured here are, from left to right, Basil Edwards, John Gordon, Mack ? (seated), DeWitt Haynes, unidentified, and Mrs. DeWitt Haynes (in car). (Courtesy of Grant County Museum.)

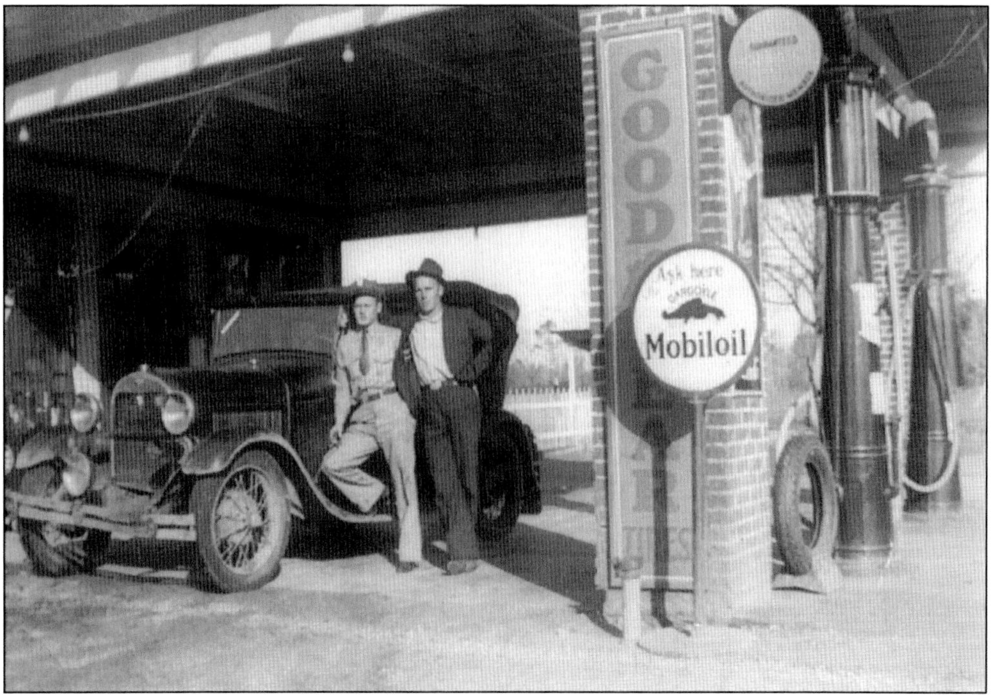


Popular Grant County citizen Okie T. Reynolds (left) and Raymond Crouse are pictured at a Sheridan service station in the 1940s. (Courtesy of Peggy Reynolds.)

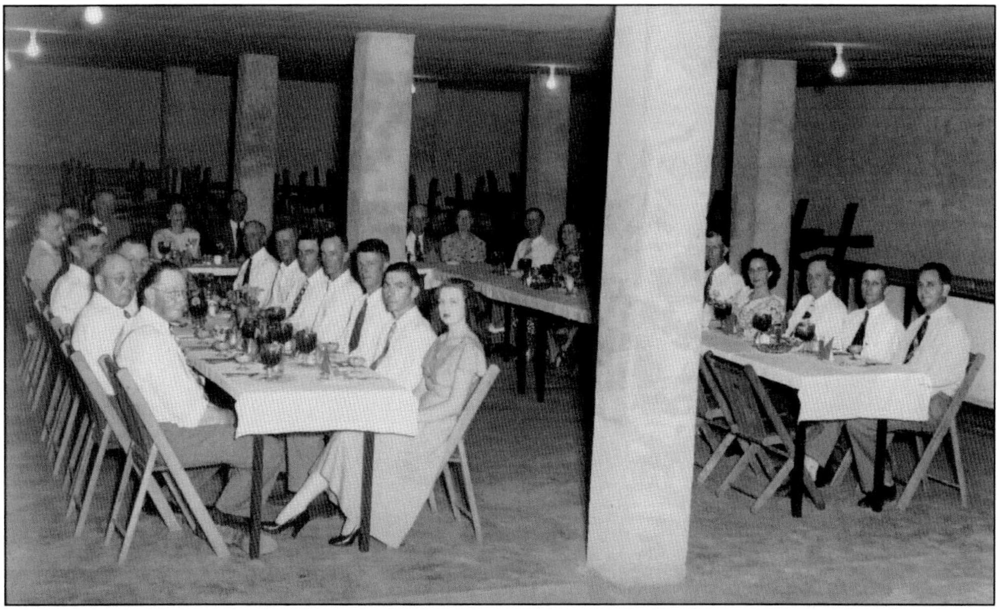

Long Bell Lumber Company honored C.E. Baxter, general manager of the Arkansas division, with a retirement banquet on July 1, 1948, in the high school cafeteria underneath McKenzie Hall. At the left table, clockwise from front left, are Tom Taylor, Warren A. Douglas, unidentified, Millard Hollinger, Coats A. Mitchell, R.N. Scales, Virgil Stuckey, two unidentified, Clay Cullins, Hilford Harrington, John Bell, and Barbara Mae Holland. In background at left, are, from left to right, C.V. Holbrook, (district manager), Kathryn Edwards, and Franklin Edwards. At right, from left to right, are C.E. Baxter, Mrs. Baxter, Taylor McRay, Mrs. McRay, unidentified, Hill Windham, Rosie Mathews, Lee Holiman, J.A. Waddell, and Curtis Monroe. (Courtesy of Grant County Museum.)

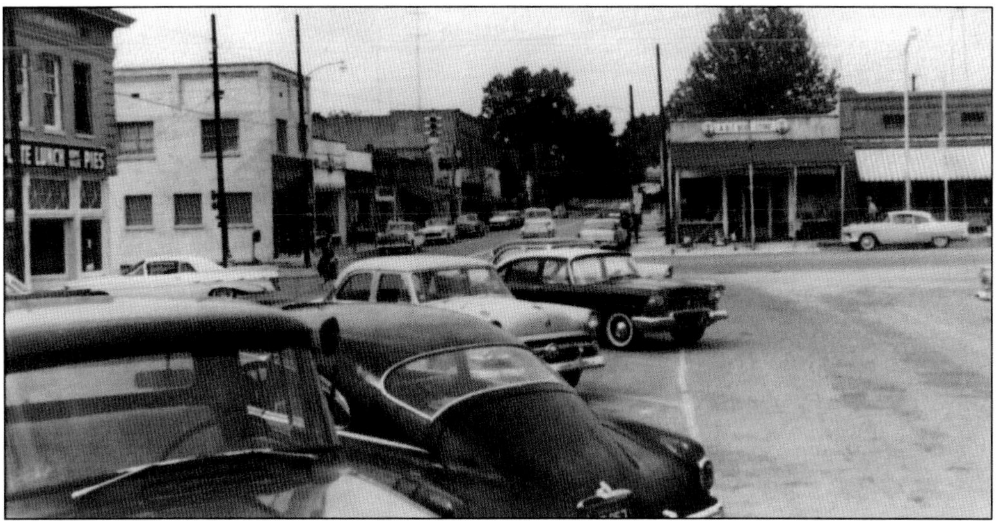


This photograph of Oak Street in 1955 demonstrates improving economic conditions. On the northeast corner of Center and Oak Streets, the ABC Bookstore, operated by Joe T. and Mary Gartman, also sold lawn mowers. At left, people could visit the Grant County Bank and the drug, grocery, or hardware stores, then see a movie at the Grant Theater. (Courtesy of Grant County Museum.)

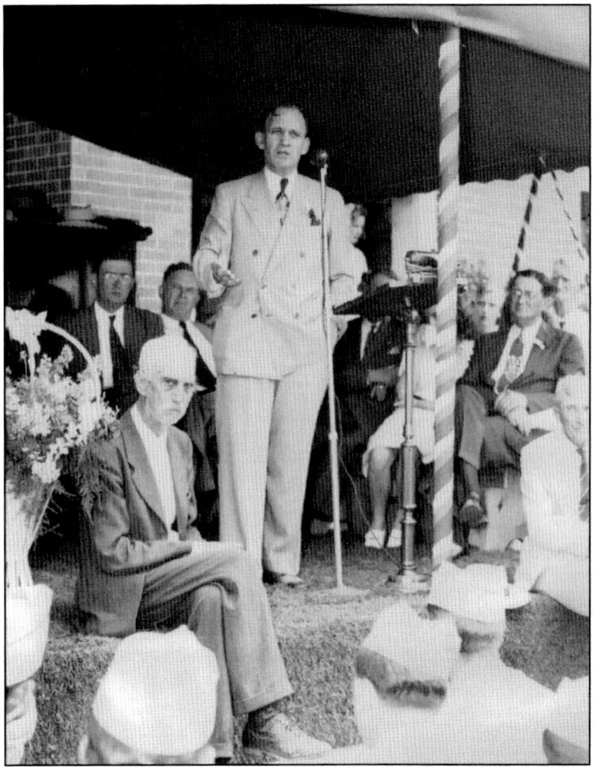

In 1946, US senator John Little McClellan spoke to hundreds at the Grant County War Memorial dedication. Senator McClellan, born in 1896 in the Ain community to Isaac and Belle Suddeth McClellan, became a lawyer in 1913 at age 17 (the youngest in the nation) by permission of the Arkansas Legislature and served in World War I in the Signal Corps. In 1934, he was elected to the US House, and he was elected to the US Senate in 1942, continuing until 1977, the longest-serving US senator in Arkansas history. His father, Isaac, is seated at front left. Behind at left is Brother Conrad N. Glover. (Courtesy of Grant County Museum.)

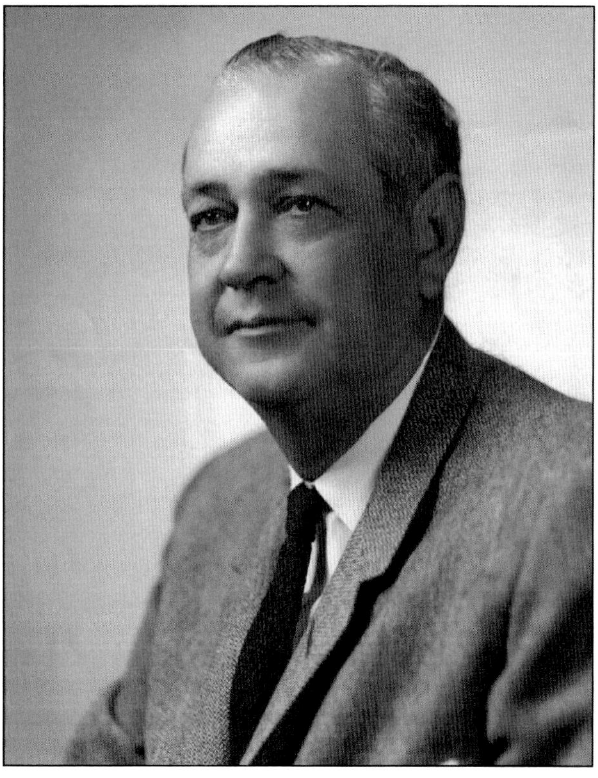

W.R. (Witt) Stephens, born in Prattsville in 1907, would develop his talent for business into a very successful investment company, Stephens, Inc., improving the lives of millions through Arkla Gas and Allied Telephone and financing the initial growth of Walmart. Stephens became the most powerful political force in Arkansas for decades after World War II. (Courtesy of Grant County Museum.)

Ray Thornton, born in 1928 to Raymond H. Thornton Sr. and Wilma Stephens Thornton, graduated from Sheridan High School, Yale University, and the University of Arkansas School of Law. His public career included serving as Arkansas attorney general, six terms in the US House of Representatives, on the Judiciary committee that proposed impeachment articles against President Nixon, as president of Arkansas State University and the University of Arkansas system, and as associate justice of the Arkansas Supreme Court. (Courtesy of Grant County Museum.)

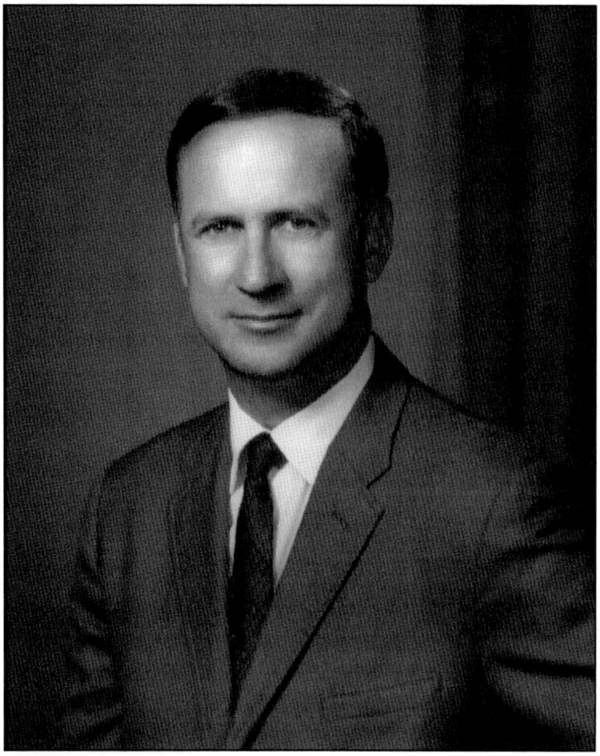

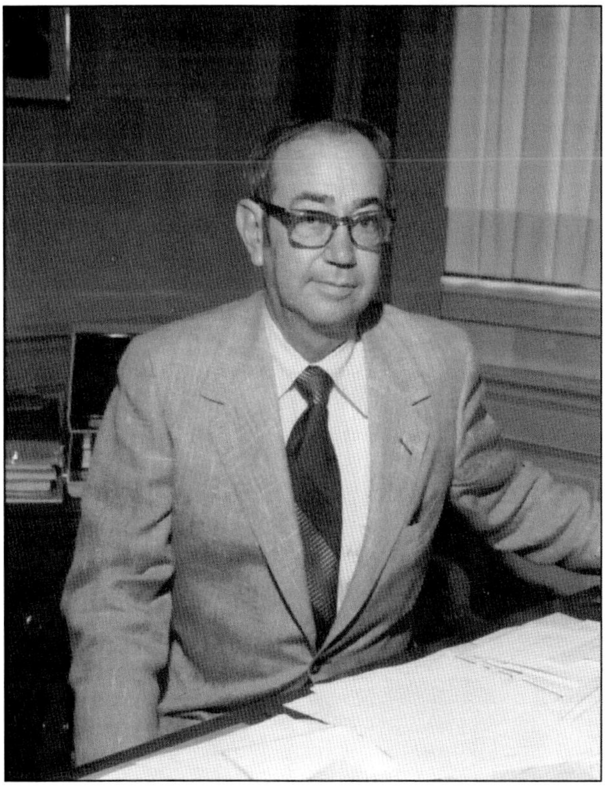

Jackson T. Stephens, born in Prattsville in 1923, helped transform economic opportunities for countless people through his partnership with his brother Witt at Stephens, Inc., in Little Rock. His philanthropy was extraordinary and continues to be appreciated by citizens of Arkansas. He was chairman of the Augusta National Golf Club, sponsor of the Masters, from 1991 to 1998. (Courtesy of Grant County Museum.)

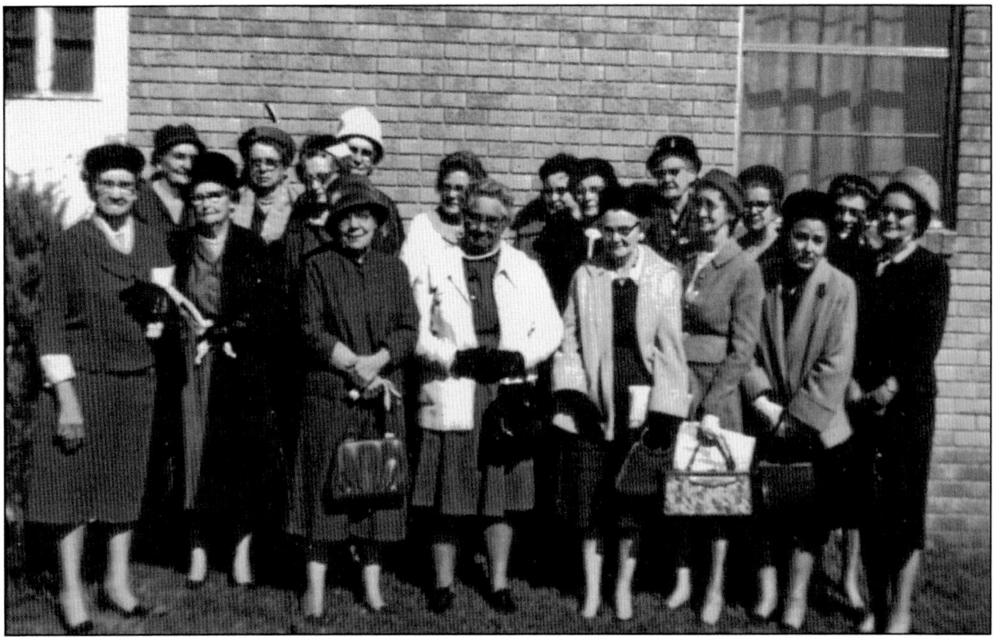

The Class of the Open Door of the Sheridan Methodist Church is pictured in 1963. The ladies are, from left to right, (first row) Mrs. M.K. Rogers, Mrs. Nelson, Mrs. Nance, Mrs. Hipp, Mrs. Blake, Mrs. Daniel, Mrs. J.H. Williams, and Mrs. O.W. Hope; (second row) Mrs. Skipper, Mrs. Jones, Mrs. Lee Harris, Mrs. House, Mrs. H. Rogers, Mrs. Pierce, Mrs. Gray, Mrs. Cotton, Mrs. Claud Williams, and Ruby Featherston. (Courtesy of Grant County Museum.)

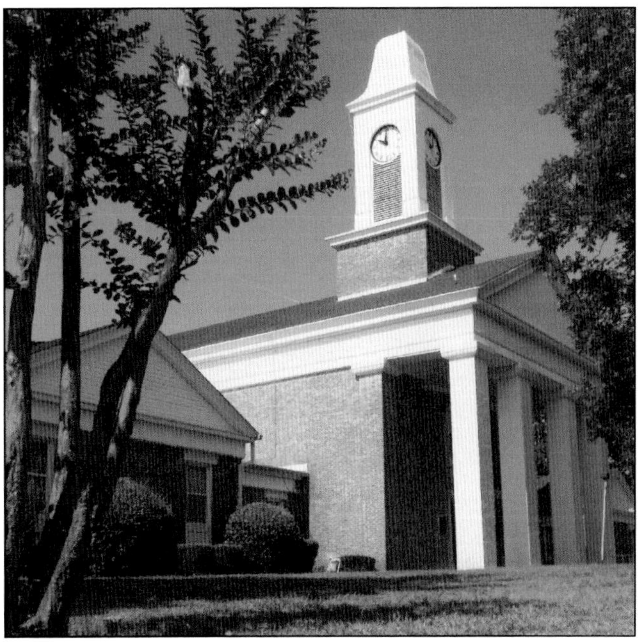

This is the fourth courthouse on the square and the second to face north. The first courthouse (burned in 1877) and the second faced Main Street to the east. The 1909 building was the first to face north toward Center Street and businesses like Glazer's Department Store, Blakely and Rexall Drug, E.A. Posey Grocery, Bruce Gartman Hardware, McCool's Café, Howard's Dime Store, Pierce Koon Jewelry, Meeks and Reid Barber Shops, Western Auto, John Gean Meat Market, J.F.H. Wilson Store, the Nixon Hotel, Lem Jones Grocery, and the Sheridan Bottling Works. (Courtesy of the author.)

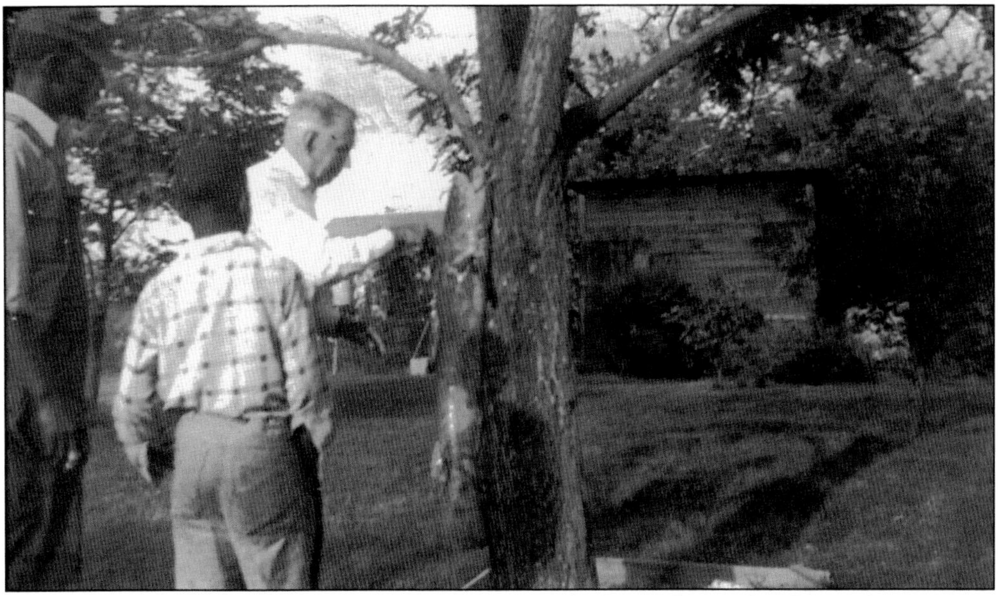

Dr. O.W. Hope demonstrates how to skin a catfish to onlookers in his yard at the northwest corner of Oak and High Streets in Sheridan. (Courtesy of Grant County Museum.)

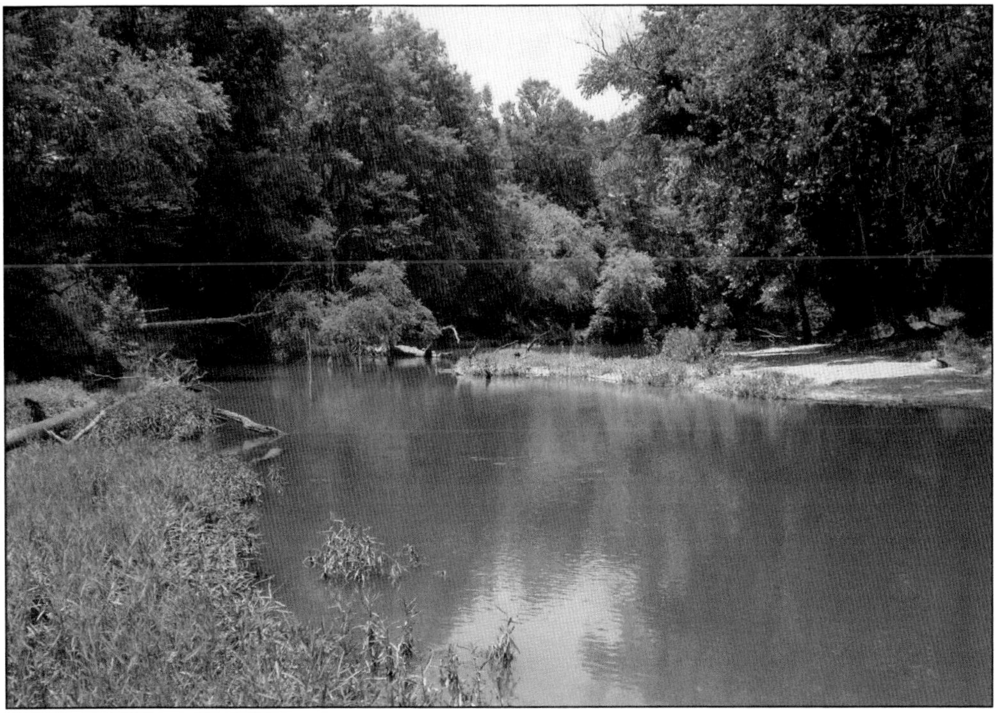

The Saline River and the opening of land in Arkansas in the 1800s offered opportunities to those who were strong and brave enough to move from eastern states. Inexpensive land once lured many westward, hoping for progress. The ancestors did as they thought correct to fulfill their destinies. Past traditions of Sheridan and Grant County are now fading into history. Much like the water of the Saline moving forward, those traditions are being replaced with ideas from newer generations striving to make their journey through time their own. (Courtesy of the author.)

Consistent with our mission to preserve history on a local level, this book was printed in South Carolina on American-made paper and manufactured entirely in the United States. Products carrying the accredited Forest Stewardship Council (FSC) label are printed on 100 percent FSC-certified paper.

MADE IN THE USA